Reproductive Decisions

MONOGRAPHS IN BEHAVIOR AND ECOLOGY

Edited by John R. Krebs and
Tim Clutton-Brock

Reproductive Decisions

An Economic Analysis of Gelada Baboon Social Strategies

R.I.M. DUNBAR

Princeton University Press
Princeton, New Jersey

Copyright © 1984 by Princeton University Press
Published by Princeton University Press, 41 William
Street, Princeton, New Jersey 08540
In the United Kingdom: Princeton University Press,
Guildford, Surrey

Library of Congress Cataloging in Publication Data will
be found on the last printed page of this book

ISBN 0-691-08360-6
0-691-08361-4 (pbk.)

This book has been composed in Linotron Times Roman

Printed in the United States of America by Princeton
University Press, Princeton, New Jersey

Contents

Acknowledgments

This book inevitably owes a great deal to a great many people, too many to mention all individually. John Crook provided the original stimulus for the project and I am grateful for his discreet guiding hand during the past decade and a half. I owe a special debt to my wife, Patsy, for her unstinting efforts in collecting the data. Among the many who made life easier and more exciting in the field, we are especially grateful to Mike Gilbert, Lionel and Joan Melville, Doris and Tom Tischler and Julius Holt and his colleagues. We are indebted to Berhanu Assefaw, Jurg Muller, and Peter Stahli (the Wardens of the Simen Mountains National Park) and to the committee and members of the Bole Valley Society for their continued cooperation and logistic support throughout two field studies. The Institute of Ethiopian Studies of the University of Addis Ababa kindly sponsored us as Visiting Scholars, while the Ethiopian Government Wildlife Conservation Organization provided logistic support and permission to work in the Simen Mountains Park. The field work was made possible by research grants from the Science and Engineering Research Council (U.K.) and the Wenner-Gren Foundation for Anthropological Research. During 1970–73, I was supported by an SERC Overseas Scholarship, and from 1977–82 by an SERC Advanced Research Fellowship. Preparation of the manuscript was assisted by a grant from the Leverhulme Trust. The field work was undertaken while I was attached to the Department of Psychology, Bristol University; most of the data analyses were carried out while I was a member of the King's College Research Centre, Cambridge; and the modeling and the writing were done while I was at the Sub-Department of Animal Behaviour, University of Cambridge. I am indebted to Professor John Brown, to the Provost and Fellows of King's College, Cambridge, and to Dr. Patrick Bateson and Professor Robert Hinde for the use of the facilities in their respective institutions. I am grateful to Professor M. Kawai for inviting me to visit the Kyoto University Primate Research Institute at Inuyama, Japan, in 1980 and to the Japan Society for the Promotion of Science for making the visit possible: the gelada story benefited greatly from discussions with Toshitake Iwamoto, Umeyo Mori and Hideyuke Ohsawa. Lorraine Weaire valiantly struggled through an endless series of drafts and revisions. Nick Davies, David Harper, and Phyllis Lee kindly commented on the manuscript, but take no blame for the errors that remain; Rose

Udics made a number of improvements to the text. Last, but by no means least, this book owes nearly everything to the 2000-odd gelada who contributed to its contents: theirs was by far the greatest contribution.

Reproductive Decisions

1 Introduction

As our knowledge of the behavior of particular species has increased with time, it has become apparent that the traditional ethological notion of "species-specific behavior" is often inappropriate for social behavior. More than anything else, field workers have come to appreciate that the degree of variability in the behavior of natural populations of animals is quite extraordinary. Concomitantly, we have seen a shift in emphasis over the past two decades from the early ethologists' view of animals responding more or less automatically to stimuli to one where animals are seen as decision-makers engaged in a process of evaluating strategic options.

This book is concerned with decision-making by animals. In it, I analyze the social behavior of gelada baboons (*Theropithecus gelada*) as a complex set of alternative strategies among which an individual has to choose. I ask the question: given that the gelada have the social system and ecological niche they do, how does an individual animal set about maximizing its personal reproductive output?

I also have an ulterior motive in that I use the gelada to illustrate a methodological approach to the study of social systems that is now beginning to yield increasing dividends. This approach takes the view that social behavior is concerned ultimately with reproduction and can most usefully be interpreted in terms of strategic decision-making aimed at maximizing an individual's contribution to its species' gene pool (see for example Daly and Wilson 1983). Of course, not everything an animal does during social interactions is immediately concerned with reproduction. Most behavior is concerned directly with objectives that are logically more proximate. For this reason, it is particularly important to distinguish between proximate and ultimate explanations of behavior. These can, perhaps, best be viewed as a series of increasingly direct influences on an individual's reproductive prospects and, at one further remove, on its contribution to the species' gene pool (see Dunbar 1983a). Reproduction, in a word, is the central problem in the life sciences, for it is the issue around which all other aspects of biology hinge.

I shall argue that such a view is becoming increasingly necessary if we are ever to understand social behavior completely. This is because it is essential to know not just what an animal does, but also what it is "trying" to do in order to understand why, in the end, it does what it does (see also Seyfarth 1980). This inevitably demands a much more sophisticated approach to the problems of social behavior: indeed, the success of such a program is dependent on the existence of a body of theory capable of making detailed predictions about behavior. That theoretical framework now exists in what has come to be known as sociobiology, while the analytical techniques have long been available in economics and operations research (McFarland and Houston 1982).

The perspective I adopt is strictly sociobiological, though I hasten to add that I do not espouse narrow "socio-genetical" arguments: the decisions made by the animals are too complex and too deeply nested in a hierarchical network to constitute a case of simple genetic determinism. Nonetheless, a broad sociobiological viewpoint is heuristically valuable in that it provides a powerful Darwinian explanatory basis for observed behavior. I take it as axiomatic that an animal's genetic inheritance obliges it to strive to maximize its contribution to its species' gene pool, but that the actual choice of means to achieve that end is a consequence of the evaluation of the relative costs and benefits of different strategies.

In this respect, I shall make frequent use of the language of conscious decision-making in defiance of Lloyd Morgan's proscription of anthropomorphisms. I do so partly because this is much the easiest way to discuss the animals' behavior, but also partly because fifteen years of field work have made it abundantly clear to me that strategy evaluation is precisely what the animals are doing (see also Kummer 1978, 1982).

Theory of Reproductive Strategies

Evolution occurs as a result of a number of processes that influence a species' gene pool. One of the most important of these is the production of offspring, since it is through offspring that an animal usually makes its main genetic contribution. Naive Darwinian considerations lead us to expect that animals will seek to maximize, by one means or another, the number of offspring they produce. As with most biological phenomena, however, sheer maximization is often counter-productive: the more offspring an individual produces, the less parental care can be given to each and the more likely it is that a high proportion of the young will fail to reach maturity. Lack (1966) gives examples from birds showing that the number of fledgling young peaks at intermediate clutch sizes. In reproductive terms, more does not always mean greater success in the long run.

Both the strategies that an animal can pursue and their relative efficiencies (in terms of generating mature offspring) are determined by the interaction of ecological, demographic, and social factors (Dunbar 1982a). The components of this system place conflicting demands on an animal's limited time and energy budgets and its morphological characteristics (Goss-Custard et al. 1972, S. Altmann 1974). Consequently, the optimal solution to one problem will often be incompatible with the preferred solutions to other problems. The resulting conflicts of interest will force the animal to re-evaluate its priorities and so to compromise on its original objectives.

The ways in which the various components of each subsystem interact are, in general, well understood. In contrast, the ways in which the subsystems themselves relate to each other have barely been touched on as yet. In particular, the important part played by demographic structure in determining the options available to an animal seems to be less widely appreciated than might be expected (Dunbar 1979a, Altmann and Altmann 1979). Demographic factors determine not only the social and reproductive opportunities available to an animal, but also the level of competition from conspecifics that it will have to face in acquiring whatever resources are relevant.

The system is also subject to density-dependent and frequency-dependent effects. These feedback effects make the "constraint-free strategy" less profitable as more individuals pursue it and are largely responsible for the generation of alternative strategies (Dunbar 1982a and references therein). By *constraint-free strategy,* I mean the strategy that would (other things being equal and in the absence of any constraints due to increased costs) be the preferred strategy in that particular socio-ecological system because it yields the highest net gain in terms of reproductive output. (In Dunbar 1982a, I refer to this by the less satisfactory terms "primary" or "normal" strategy for the species.) Note that a constraint-free strategy is not the same as an "ideal free strategy": as originally defined by Fretwell (1972), the ideal free distribution is that to which the population evolves (in a non-Darwinian sense) once the frequency-dependent and other constraints are imposed on the constraint-free strategy.

Within the context of the constraints imposed by these factors, animals can choose among a range of strategies. The degree to which the set of strategies is stable in an evolutionary sense depends on the extent to which their profitabilities equilibrate in the long term: that is to say, on the extent to which they yield similar numbers of offspring (or genes) when summed over a lifetime (see Dunbar 1982a, 1983a).

In general, an animal may be expected to pursue those options that are most profitable to it, where profitability may be measured in terms of, for example, expected lifetime reproductive output. Of course, animals do not assess the numbers of offspring that any given strategy is worth: that would require an absurd degree of sophistication even for humans. Rather, they base their decisions on more proximate cues that, over evolutionary time, have come to be correlated with lifetime reproductive output. These proximate cues can take a variety of forms, ranging from overt events (such as the number of matings or mates acquired per unit time) to less easily quantified psychological values (such as general feelings of "contentment" or security). McFarland and Houston (1982) refer to the decision rules based on these cues as "rules of thumb." The correlations between these cues and ultimate profitability (measured in terms of genes contributed to future generations) are rarely one-to-one (Dunbar 1982b). Consequently, we cannot necessarily expect individual animals to make decisions that are evolutionarily optimal, though we can expect them to make decisions that are within a degree of latitude of those optimal decisions on the average.

Reproductive Strategies of Gelada Baboons

My immediate concerns in this book are (1) to describe the range of reproductive strategies pursued by gelada baboons, (2) to identify the proximate factors that give rise to these strategies and, where possible, (3) to evaluate their relative efficiencies.

As is well known, gelada reproductive units contain only one breeding male, and this defines and limits the range of reproductive strategies open to individuals of either sex. Male strategies are mainly related to methods of acquiring control over reproductive units. Females, on the other hand, face a more diffuse set of problems, and the range of strategies open to them is in consequence both more closely tied to their social relationships and less easily discerned by the observer. Thus, as is often the case, the problems faced by males and females are quite different, and the optimal solutions they would prefer are commonly in direct conflict with each other's interests. Part of my task here will be to determine how these strategy sets interrelate in order to understand how one sex's options limit the other sex's behavior.

To be able to do this, we need to know a great deal about the animals' background biology, both ecological and social. As far as the gelada are concerned, most of the relevant information has now been published in monographs (Dunbar and Dunbar 1975, Kawai 1979a) and an extensive series of papers. Because few people will be familiar with all these publications, I take the liberty of summarizing the most relevant details from this literature in the first few chapters. In doing so, I have tried to avoid providing a general overview of gelada biology. Instead, I have concentrated on those aspects that bear directly on the animals' reproductive strategies. Without this information, the naive reader is apt to raise all sorts of obviously inappropriate alternative explanations for particular phenomena. Those who require more detailed discussions are referred to the original sources cited in the text. This is especially important with respect to many of the causal statements in these chapters: these will often seem to be based on correlations, when reference to the original sources will reveal that the causal inferences are based on very much more detailed logical and evidential analyses.

Chapters 6 through 15 constitute the meat of the book and present both new data and previously unpublished analyses relating to gelada reproductive strategies. The first four of these chapters deal with female strategies, the remainder with those of the males. The inferential process will generally be very much more explicit here. Finally, in Chapter 16, I reconsider certain theoretical issues in the light of these analyses.

It should be noted that I make no attempt to evaluate the adaptive significance of the gelada social system. Rather, I am concerned with just one component of that system, namely, reproductive strategies within the constraints imposed by a social system that is assumed to have been determined by other factors. The general form of the social system and the species' ecological niche can be considered as constraints within which the individual animals make their decisions, even though in reality it is a two-way process. For present purposes, we can assume an explanation for the system's evolution along the lines suggested by the classical socio-ecological literature (see, for example, Crook and Gartlan 1966, Crook 1970, Denham 1971, Goss-Custard et al. 1972), even though these explanations are almost certainly incorrect.

One other point needs to be made explicit. Genetic evolution is a consequence of fitness, a population genetic concept defined in terms of selective advantage (i.e. the rate at which an allele spreads in a population relative to the rates of spread of other alleles at the same locus: see references quoted in Dunbar 1982b). In practice, we are invariably obliged to use more easily quantified measures such as reproductive success, even though the relationship between these measures and fitness itself is not necessarily one-to-one. For practical reasons, I shall in general assume that lifetime reproductive output is a sensitive index of fitness (see also Grafen 1972) unless a particular context forces me to do otherwise.

Most of the data on which this volume is based were obtained from gelada living in the Sankaber area of the Simen* Mountains National Park in northern Ethiopia during field studies in 1971–72 and 1974–75. Additional data derive from the Bole Valley some 500 km to the south (based on field work in 1972 and 1974) and from the Gich area of the Simen Mountains (based on our own brief study there in 1971 and the more extensive project carried out by M. Kawai and co-workers in 1973–74). Detailed descriptions of the study areas can be found in Dunbar and Dunbar (1974a, 1975) and Kawai (1979a).

The data themselves derive from three main sources.

The demographic data were obtained from repeated censuses of the study populations. All the members of 11 of the 31 reproductive units of the two main bands in 1971–72 and of 15 of the 17 units in the 1974–75 study were known individually, at least within the context of their particular units. A number of other adults in the remaining units were individually recognizable under any circumstances. Almost every unit in the population (five bands in each study) could be instantly identified, either from its composition or by individually recognizable members. A detailed discussion of demographic methods can be found in Dunbar and Dunbar (1975) and Dunbar (1980a). Terminology and symbols for demographic variables follow standard practice (e.g. Caughley 1977).

The data on the structure of social relationships within units (especially those discussed in Chapter 10) derive from detailed studies of individual units. Scan censuses of non-agonistic interactions were used to determine the overall pattern of social relationships for 11 units in 1971–72 and 14 units in 1974–75. In order to standardize the time base, sampling was carried out on a whole unit as long as there was at least one dyad interacting. This time base is referred to as *potential social time*. In the gelada, most social activity is confined to the first and last hours of the day, with the period between 1000 hrs and 1700 hrs being devoted more or less continuously to feeding (see Dunbar 1977a, Iwamoto 1979). Most of the data derive from the morning and evening

---

* Those familiar with earlier publications on the Simen gelada will notice differences in the spelling of place names. *Geech*, for example, is now spelt *Gich*, while what used to be *Simien* is now *Simen*. Our original spellings followed those in use in 1971–72. A standard English–Amharic transliteration system (Lewis 1959) has recently been introduced (see Stahli and Zurbuchen 1978) and it now seems preferable to use this.

social periods. More detailed sampling of interactions was carried out for 11 of the units during the 1974–75 study: these were used to study dominance relationships, coalition formation, oestrous behavior and the behavioral bases of social relationships. Details of the methods and sample sizes can be found in Dunbar (1980b, 1983c).

The final set of data concerns events associated with the reproductive strategies themselves, especially those of the males. These include events (such as takeover fights and entry by males into units as followers) that occur rarely and at unpredictable intervals. It is impossible to sample these events and the behavior that occurs during them other than by opportunistic recording of all observed occurrences (i.e. by *ad libitum* sampling, *sensu* J. Altmann 1974). Usually, once such an event had been detected, the unit was intensively sampled for 2–3 hours each day until things had settled down again. This usually involved scan-sampling of all interacting dyads and *ad libitum* recording of all inter-actions involving the main individuals concerned.

A great emphasis is placed on female kin relationships throughout this book. Without the benefit of a very long-term study, it is usually not possible to know which females are related. However, as with many other species of Old World monkeys, gelada females remain in their natal units and do not normally desert them. Analyses of the pattern of grooming interactions suggest that closely related females (mothers, daughters, sisters) form strong grooming partnerships; these were therefore used to determine matrilineal relationships (see Dunbar 1979b). Walters (1981) has shown that grooming frequencies are a good indi-cator of known mother-daughter relationships in *Papio cynocephalus*. This was subsequently shown to be true of a small group of captive gelada as well (Dunbar 1982c). There seems little reason to doubt the validity of the mother-daughter assignations we have made for the San-kaber gelada: to do so would require a set of even more complex hy-potheses to account for the peculiar age distributions observed in the grooming dyads at Sankaber (and, incidentally, at Gich and else-where). In this book, the term *matriline* refers to the set of females whose extension is defined by the reproductively mature female des-cendants of a living female (or one that has only recently died). It does not refer to the extended lineages of the kind found in the literature on *Macaca mulatta* and *M. fuscata,* where the founding matriarchs were generally defined as the set of adult females present in the group when studies were first started two to three decades ago (most of whom are now long since dead). Thus, while matrilines in the macaque literature often refer to sets of 20 or 30 females of all age classes (only half of

Figure 1   Observation conditions at Sankaber.

whom may still be living), in this book the term refers to sets of 1–6 (average 2) living reproductive females.

Dominance relationships among the members of individual reproductive units were determined from analyses of wins and losses in approach-retreat encounters that occurred in non-social contexts. The members of each unit could be ranked in a simple linear hierarchy on the basis of these data (for details, see Dunbar 1980b). It should be noted that dominance relationships between members of *different* reproductive units are in general irrelevant to the present story and are not considered here.

The detail into which we will be able to go in unfolding the gelada story has been possible because of two important factors. First, the unique structure of gelada society (small reproductive units that associate closely in large herds) made it possible to obtain unusually large sample sizes very rapidly. Not only could a very large number of individuals be observed and censused regularly, but the exceptional observation conditions made it possible to do this with minimum effort. We could expect to see 15–25 reproductive units comprising 250–350 animals on any given day (Fig. 1). This made it more likely that not only would we see rare events, but that we would see enough of them to have a statistically useful sample. Second, the fact that we could observe and sample the behavior of a large number of units meant that we could do something that has never been done before, namely, ana-lyze the social structure of those units on a quantitative basis in relation to their demographic structure. This has provided profound insights into the dynamics of the gelada social system that would otherwise probably not have been possible without at least two decades of con-tinuous field work. The importance of this factor in the analyses that follow cannot be overestimated; it shows how valuable the comparative method can be when applied across groups and populations of the same species. Of course, the gelada are in many logistic respects unique. Nonetheless, even the comparisons that we have made for as few as three populations have yielded invaluable insights into the dynamics of the species' ecology and demography (although this has been possible with such a small sample only because the populations differed mark-edly on the environmental variables under consideration).

Definitions

Definitions of the terms used for social units and the various levels of gelada society are given in the next chapter. I here define the terms used to refer to types of animals within these units. (For more detail, see Kawai et al. 1983.)

Males who "own" harems of females (in that they associate contin-uously with and have exclusive mating access to those females) are termed *harem-holders* (or just harem males). Other adult males who associate regularly with a reproductive unit (but do not, in general, have sexual access to the females) are termed *followers*. Followers may be either young adult males or old males well past their prime (see Dunbar and Dunbar 1975). Terms for the age classes for each sex are defined in Kawai et al. (1983) (see also Dunbar 1980a). Males are con-sidered to be reproductively mature at 6 years of age (though they undergo puberty at the age of 3–4 years); they continue to put on weight until 8–9 years old, which is taken to be the dividing line be-tween young and old adults. Females are considered to be reproduc-tively mature at puberty (ca. 3 years of age): they are classed as juveniles until they are 4 years old, as subadults between 4 and 6 years of age, as young adults from the ages of 6 (when they complete physical growth) to 8, and as old adults thereafter. (Females more than 11 years old are sometimes distinguished as very old adults.) The female's paracallosal skin, normally slate grey in immature animals, turns purplish at pu-berty; it remains this color until the female is about 6 years old, when it gradually begins to turn pink, a process that takes 18–24 months (see

Dunbar 1977b). Thus, the color of a female's paracallosal skin provides a reliable guide to her approximate age.

In the chapters that follow, I distinguish between the total number of animals of all ages and both sexes in a reproductive unit and the number of reproductive females in that unit. I refer to the first as *unit size* and the second as *harem size*. *Reproductive females* refers to all post-puberty females, whereas *mature females* refers to all females older than 4 years of age.

Pairs of individuals who spend more than 10% of their potential social time interacting with each other are referred to as *grooming partners*, the dyad they form being termed a *grooming dyad* (see Dunbar 1983b).

To avoid confusion over the names of reproductive units, the units censused in the two studies were given different prefixes: 1971–72 units have the letter *H* (e.g. H21), while 1974–75 units have *N* (e.g. N21). Owing to demographic changes between the two studies (notably fissions and takeovers), only a small proportion of the units observed in the first study were genealogically the same as those observed in the second study. Consequently, the identification numbers following the study year prefixes are quite independent: N21 is not the same unit as H21 two years later. Although some of the 1974–75 units were known (from the presence of identifiable members) to be specific units from the 1971–72 study, the proportion of known identities was too small to make the use of the same numbering system worthwhile.

Use of Modeling

The function of modeling in all sciences is to allow complex systems to be studied by isolating the minimum set of variables that describes the behavior of that system. Within this framework, modeling can take a variety of forms and can be used for several, quite different, purposes. I use modeling (1) to generate null hypotheses against which to evaluate observed distributions, (2) to study the long-term consequences of particular behavioral strategies, and (3) to determine the relative importance of the variables that give rise to particular patterns of behavior. The second and third are often two sides of the same coin and, in some cases, involve fairly complex models.

One of the key problems in studying the functional and evolutionary aspects of the behavior of large, relatively long-lived mammals is that direct estimates of the consequences of different behavioral strategies in terms of lifetime reproductive output can seldom be obtained. In the case of the gelada, for example, such a study would require about 20 years, while chimpanzees or gorillas would require studies on the order of 30–40 years to ensure that all the members of one cohort were followed through a complete lifetime (see Teleki et al. 1976). However, in those cases where we know enough about the animals under study to be able both to specify the relationships between all relevant variables *and* to quantify these relationships, we can use modeling to study the consequences of behavior under circumstances where we have no prospect of being able to carry out proper empirical studies.

If we do not do this, we must either accept that our knowledge of the world about us will always remain incomplete or we must use simpler, more short-lived species as models of larger, less easily studied taxa. Neither strategy is an encouraging prospect, least of all the second because of the number of assumptions involved in deducing conclusions about the behavior of one species from the behavior of another. Each species is the unique outcome of a particular set of biological and environmental interactions over time; consequently, one species' responses to a given set of environmental conditions will often be quite different from those of another (see Dunbar 1984a). These will, in turn, impose quite different constraints and selection pressures on the animals' reproductive and social behavior.

This is often assumed by the philosophically naive to mean that the life sciences have no universal principles and therefore lack predictive power. This, however, is to confuse different levels of explanation, and it is one of the fundamental assumptions of this book that such principles do exist. Biological problems are inherently complex because they involve the interaction of many component elements, each with its own predictive universal principle. Consequently, the key to understanding a species' behavioral biology lies in being able both to identify all these components and to evaluate their relative influences on the more general problem. Only then is it possible to make valid predictions as to what behavior is the most likely to evolve. In most cases, simulation modeling is the only means of handling such a complex network of causal relationships. The models I use here generally take the form of multivariate dynamic models of the kinds commonly found in economic and operations research contexts (see, for example, Thierauf and Klemp 1975, Jeffers 1978). The application of this approach to social behavior constitutes a powerful analytical tool that will permit rapid advances in our understanding of social systems at both the individual and the socioecological levels (see also Nagel 1979).

One final important advantage of modeling should be mentioned. In seeking to understand and explain behavior, we need to know not just how to predict it, but also why a particular prediction occurs (Gale 1979). This requires us to go beyond a mere curve-fitting exercise to determine precisely why it is that any two variables are related in the way they are. Modeling forces us to axiomatize the network of hypotheses, and so to specify precise quantitative relationships (see Maynard Smith [1978], who stresses the importance of stating *all* assumptions clearly when modeling). Once a system has been modeled, sensitivity analysis can be used to determine which relationships contribute most to the predicted outcome. Sensitivity analyses are often used to demonstrate that models are robust—i.e. that the outcome occurs regardless of the precise shape of the quantitative relationships between the variables. This is a useful exercise for simple systems, but for systems that are inherently variable (and in which the variability itself is the object of study), it may tell us little we do not already know. What it can do, however, is tell us how important each variable is to the outcome. I shall make as much use of this second approach to sensitivity analysis as I shall of the first.

Statistical Analyses

The statistical procedures used in this book are taken from standard sources (usually Siegel 1956, Sokal and Rolf 1969, or Pollard 1977). Two-tailed tests are used throughout, except where specific hypotheses are being tested. I treat significance levels (i.e. *p*-values) as estimates of confidence in a Bayesian sense (following Salmon 1966, Lindley 1970, and others). This means that I tend to specify *p*-values rather more precisely than is normally considered necessary and that I do not accord the value $p = 0.05$ the magical property that is commonly associated with it. Rather, my belief in any one hypothesis is a simple function of the *p*-value associated with its statistical test, the more so when compared to the *p*-values obtained for alternative explanations of the same phenomenon. This should not be seen as an attempt to eke out significant results from dubious data, but rather as an attempt to place a more satisfactory interpretation on the meaning of statistical tests and the inferences drawn from them. Nonetheless, in the interests of minimizing confusion, I will use the term *statistically significant* in its conventional sense with reference to the $p = 0.05$ level.

It is worth observing here that a Bayesian approach to inference using multiple-hypothesis-testing is a powerful means of strengthening evolutionary and functional explanations of behavior (see also Platt 1964). A functional perspective itself is heuristically valuable because it forces us to find an explanation for a phenomenon, thereby making us think more deeply about it. However, because functional explanations can easily degenerate to the point where they have no greater intrinsic justification than their own null hypotheses, it is essential that they be carefully evaluated and tested. Unfortunately, it is often not possible to carry out rigorous tests with observational studies. This problem can usually be circumvented by ensuring (1) that the explanation is consistent with as many other known aspects of the system as possible, (2) that all alternative explanations can be excluded, and (3) that a convincing physiological or behavioral mechanism can be shown to underlie the functional explanation.

In a number of cases where samples are too small to generate meaningful estimates of significance, a procedure due to R. A. Fisher can be used to pool the results of a number of independent tests of the same hypothesis (see Sokal and Rolf 1969, pp. 621–624). This allows us to determine how likely it is that a distribution of p-values as extreme as those observed would be obtained by chance. This procedure is particularly valuable when the proper unit of analysis is the set of individuals within a harem: in such a case, the sample size is often only 3–5 animals, but data may be available from as many as 10 such units.

2

Structure of Gelada Populations

The gelada social system is a complex arrangement of hierarchically organized social groupings, each of which corresponds to a different functional unit. These groupings are analogous to those of the hamadryas baboon, *Papio hamadryas* (see Sigg et al. 1982), these two species apparently being unique among the primates in the degree of organizational complexity that they have evolved.

In this chapter, I describe the social units that make up the gelada system. The species' ecological niche and population dynamics will be outlined in Chapters 3 and 4, while the structure of social relationships between the individual animals will be discussed in Chapter 5.

Structure of Gelada Society

Traditionally, the basic social unit of the gelada is considered to be the one-male group (Crook 1966), and this is certainly the smallest independent grouping within the social system. These units, together with all-male groups of bachelor males, associate together in a series of higher-level groupings. Kawai et al. (1983) have viewed these patterns of association between units as ascending levels of clustering on a tree-dendrogram. Figure 2 shows this for a nominal group of units. The variable on the y-axis can be either of two equally good indices: (1) the probability of any two units being together on any given day or (2) the correlation between these two units' use of the habitat (i.e. the similarity in their ranging patterns). These association patterns, however viewed, are highly variable and, strictly speaking, they form a continuum over the unit probability range. However, the tendency for association probabilities to cluster at specific levels allows us to define groupings that possess some real significance for the animals themselves. The animals probably view the situation in terms of units whose members are more or less familiar to them, familiarity being a consequence of frequency of association. Frequency of association correlates strongly with similarity of home range, though whether units have similar home ranges because they associate together frequently or associ-

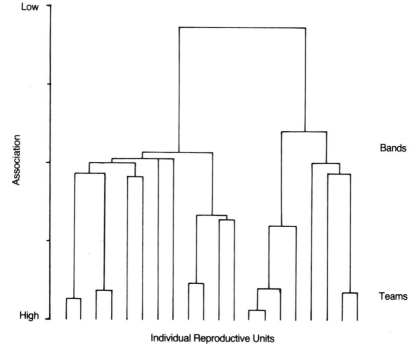

Figure 2   Cluster-analysis of association patterns among a nominal group of reproductive units belonging to two bands. The degree of association between any two units can be estimated either by the frequency with which they occur in the same herd or by the correlation between their respective ranging patterns. Tendencies for distinct clusters to form at specific levels of association reflect the grouping patterns within gelada society. These are identified on the right-hand side of the dendrogram. (Redrawn from Kawai et al. 1983, Fig. 1.)

ate together because they share a common range is a moot point. I am inclined to favor the view that units share a common range because they associate with each other frequently, and that they associate frequently because they are the product of the successive fission of units in the past (Dunbar and Dunbar 1975; see also Kawai et al. 1983).

The two main components of the system are the individual reproductive units and the clusters of units (termed *bands*) that share a common home range. The one-male units provide the context in which most social behavior and all reproductive activities take place, whereas the band is the basic *ecological* unit (being precisely analogous to the *Papio hamadryas* band). The band is a multimale unit that is ecologically equivalent to the typical *Papio* and *Macaca* troop. Studies by Sho-

Figure 3   Part of a herd of gelada baboons.

take (1980) suggest that the gelada band may also be the fundamental *genetic* unit in that members of a given band are genetically more homologous than members of different bands. Shotake's calculations indicate that gene flow between neighboring bands may be as little as 5% per generation, an estimate that accords reasonably well with observed migration rates. Although bands do exchange individuals (and genes), such exchanges are probably most likely to occur between bands whose units associate together frequently.

Some units of a band associate together more closely than they do with other units of that band, suggesting that it may be possible to identify a social grouping (termed a *team*) intermediate between the reproductive units and the band (Kawai et al. 1983). Teams are probably the product of the recent fission of a larger unit where residual social bonds between the members of the two units continue to keep them together (see Dunbar and Dunbar 1975, pp. 103–106). Teams tend to be characterized by the fact that they are invariably found together and that members of the two units will often respond to each other's contact calls (something that is exceptionally rare between units that associate less frequently).

The last level of grouping that we need to consider is the *herd* (Fig. 3). Although Kawai and co-workers (Kawai 1979a) used this term to refer to bands, it is now agreed that the term *herd* should be used only to refer to unstructured consociations of units of no temporal stability,

Figure 4   Two reproductive units feeding next to each other. The males clearly define the centers of the two units, although the edges of the units merge.

as originally defined by Crook (1966). The term *herd* is thus used in exactly the same sense as the term *party* was in the older primate literature and refers to units who happen, at any particular moment, to be in the same place. A herd is unstable and its constituent units may come and go more or less at will. A herd may consist of one or more units of one band associating with one or more units of any other neighboring band (or bands).

Structure of Social Units

REPRODUCTIVE UNITS

Gelada reproductive units (Fig. 4) are highly variable in size and composition. Units as large as 50 individuals have been recorded in areas where hunting pressure is high, though in undisturbed habitats units larger than 20 individuals are rare.

The typical reproductive unit contains a single breeding male and an average of four (range 1–10) reproductive females, together with their dependent young. A proportion of the units in each band contain additional adult males, termed *followers*. In general, followers are either young or very old males and do not have mating access to the females in the unit. The only exception is that a *young* follower may in some

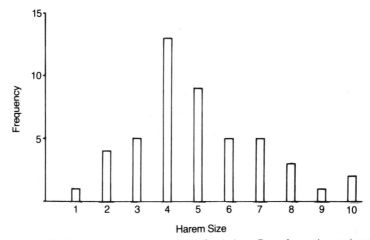

Figure 5 Distribution of harem sizes at Sankaber. Data from the main study bands of both field studies are pooled.

cases have strong social *and* sexual relationships with one (exceptionally two) of the females in his unit. Although most units have only one follower (either young *or* old), as many as five mature males have been recorded as followers of a single unit.

Table 1 summarizes the main demographic features of reproductive units in the three areas for which there are detailed data. The distribution of the number of reproductive females per unit for the main study bands at Sankaber is shown in Figure 5.

ALL-MALE GROUPS

Young males leave their natal units as juveniles or young subadults to join or form all-male groups (Dunbar and Dunbar 1975). All-male groups are far more labile in their attachment to their natal bands and may spend a considerable proportion of their time wandering alone or attached to an adjacent band. The mean size of all-male groups in the three study areas was 6.5 (range 2–15, $n = 28$).

The processes whereby all-male groups are formed are poorly understood, though Dunbar and Dunbar (1975) suggested that young males might become members of an all-male group either by joining an existing group (perhaps one to which close relatives already belong) or by forming a new group through the gradual intensification of the bonds among members of the more transitory juvenile male peer groups.

Some light can be thrown on these hypotheses in a rather simple way. Figure 6 compares the variance in the ages of males in 18 San-

Table 1

Demographic characteristics of reproductive units in the three study areas.

	Mean Composition				Total Size		
Study Area	Adult males	Subadult males	Adult females	Immatures	Mean	Range	n
Bole	1.1	0.6	5.9	9.5	17.1	8–28	10
Sankaber[a]	1.3	0.1	4.1	6.5	12.0	3–26	48
Gich	1.2	0.3	3.9	4.4	9.9	2–17	31

[a] Data for the two field studies pooled (main study bands only).

Table 2

Demographic characteristics of bands in the three study areas.

	Total Size		Mean number of:		Adult sex ratio[a]	Males in all-male groups (%)	Number of bands	Multi-male units (%)	Immatures per female
Study Area	Mean	Range	All-male groups	Harems					
Bole	60.3	48–78	1.3	3.3	4.21	28.6	3	20.0	4.20
Sankaber[b]	131.5	30–262	1.5	10.7	2.75	10.1	11	38.5	2.75
Gich	107.2	27–170	1.5	9.7	2.35	16.9	6	22.8	2.37

[a] Adult females per adult male.
[b] Data for the two field studies pooled.

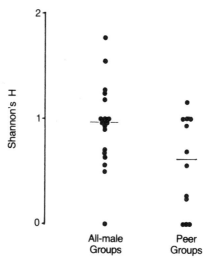

Figure 6   Variance in the age structure of 18 all-male groups and 12 male peer groups at Sankaber, using the Shannon–Wiener index of diversity, H, as a measure of variance.

kaber all-male groups with that for 12 juvenile male peer groups sampled at random at Sankaber in 1971–72. The Shannon–Wiener index of diversity, H, is used as an index of the variance in composition. The variance is greater in all-male groups than in peer groups (mean H = 0.897 vs. H = 0.587, respectively; Mann-Whitney test z = 1.757, p = 0.078 2-tailed), suggesting that it is unlikely that all-male groups often develop out of relationships existing in juvenile peer groups alone. Rather, most juvenile and subadult males probably attach themselves to existing all-male groups.

BANDS

Bands are highly variable in size, the variance within study areas being far greater than the variance in mean band size between habitats. Band size at Sankaber ranged from 30 to 262 animals (n = 11 bands), while comparable ranges were observed in other areas (Table 2). The average band contained 9.8 reproductive units and 1.4 all-male groups, comprising a total of 113.5 individuals (n = 20 bands at Sankaber, Gich, and Bole).

Ohsawa and Dunbar (1984) found that key demographic variables such as adult sex ratio, the proportion of breeding males, the percentage of multimale units, and the ratio of immatures to females varied considerably both between bands in the same and different habitats and within the same band in different years. Within-band variation was often as great as between-band variation. Taking overall averages is consequently somewhat deceptive, since it implies too static a picture; nonetheless, Table 2 does give some impression of the order of magnitude with which we are dealing.

3 Ecological Constraints

In this chapter, I provide an overview of the more important aspects of gelada ecology. I draw on comparative data to show how the species' dietary specializations influence time budgets, population density, and ranging patterns. This specialization has important consequences for the animals' reproductive strategies since it limits and constrains their day-to-day behavior. Detailed accounts of the species' ecology can be found in Dunbar (1977a), Iwamoto (1979), Kawai and Iwamoto (1979), and Iwamoto and Dunbar (1983). These sources should be consulted for the bases on which causal relationships were deduced from comparative data.

The Theropithecine Niche

DIETARY SPECIALIZATION

Gelada rarely eat anything other than grass products (Fig. 7). Given the choice, they show a clear preference for grass blades, but, if these are in short supply (as may be the case during the dry season), they will dig for roots and rhizomes (Fig. 8). It is worth observing that this shift to underground food sources occurs mainly in areas where there is heavy overgrazing by domestic stock (see Iwamoto and Dunbar 1983).

MORPHOLOGICAL SPECIALIZATIONS FOR GRAMINIVORY

The gelada's dietary specialization is reflected in several important morphological characteristics. These illustrate the species' total commitment to an ecological niche that is unique among the primates (see Jolly 1972, Dunbar 1984a).

Grasses require little biting and a lot of chewing before ingestion, with consequent heavy wear on the molar teeth. The theropithecines as a group are characterized by small incisors and greatly enlarged molars (see Jolly 1972, Szalay and Delson 1979). The molar teeth are also hypsodont, the enamel layer being both deep and heavily ridged. These are adaptations designed to prolong the life of the toothrow, as is the case with many other grazing mammals.

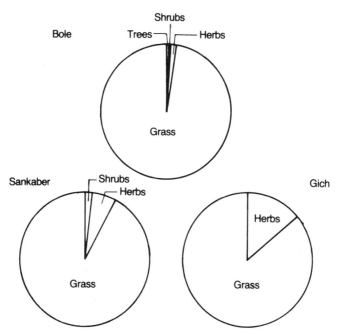

Figure 7 Composition of gelada diets in three study areas, based on scan samples of feeding animals (from data given by Iwamoto and Dunbar 1983).

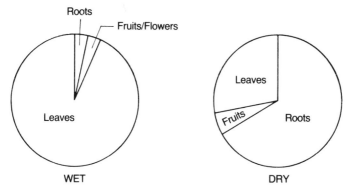

Figure 8 Comparison of wet and dry season diets of the Sankaber gelada, showing the seasonal shift from grass leaves to grass roots.

Gelada are able to minimize the wear on the teeth from grit ingested with the grasses by being highly selective when feeding. They have the highest "opposability" index of the nonhuman primates (a measure of the extent to which thumb and index finger can be opposed: Napier and Napier 1967). This allows them to select green grass blades very skillfully from even the driest sward (Dunbar 1977a).

Gelada also have the highest "robusticity" index of any nonhuman primate (defined as the diameter of the fingers relative to their length: Jolly 1972). Their relatively short, stubby fingers make it possible for them to dig for roots and bulbs more efficiently and rapidly than other monkeys. This particular adaptation was probably crucial in allowing the genus to invade the well-established grazing ecosystem of the African Pliocene grasslands, for it permitted the animals to tap a food source that was not available to the grazing ungulates (Dunbar 1976, 1984a).

A fourth adaptation relates to the stance adopted by the gelada when feeding. Typically, they feed in a sitting position, whereas *Papio* baboons tend to stand tripedally (Dunbar 1984a). Gelada often sit on their haunches rather than directly on the ground and use a unique shuffling gait when the distance to be moved to a new feeding site is very small (Wrangham 1980). Up to 30% of the distance traveled each day is covered by shuffling. This tendency to feed in a seated position may explain why the gelada possess a sexual skin on the chest as well as on the perineal area (Fig. 9): the latter would obviously be of little signal value when the animal spends such a large proportion of its time seated.

Finally, the gelada have become almost completely terrestrial and have one of the highest "terrestriality" indices (ratio of foot width to length) of the primates (Jolly 1970b). Gelada rarely climb into trees, even when there is ample opportunity to do so. Of 13,755 animals censused in scan samples at Sankaber, only 55 (0.4%) were in trees or bushes. About the same proportion of feeding animals were physically off the ground (Dunbar 1977a).

A PROBLEM IN DIGESTIBILITY

Herbivores face a serious problem in digesting their food. The cell walls of plants contain a large amount of relatively indigestible cellulose that greatly reduces the efficiency with which nutrients can be extracted from the ingesta. Herbivores have evolved a number of strategies to cope with this problem. Among the most familiar are rumination (a characteristic of the ungulates, though foregut fermentation has independently evolved in certain of the folivorous primates) and coprophagy (where the ingesta are passed twice through the entire digestive system: a strategy adopted by rabbits and hares).

Figure 9   Sex skins of the female gelada baboon. (a) A young female's chest, showing oestrous vesicles around the edge of the sex skin. Note the centrally placed teats: in multiparous females, the teats hang in the mid-line. (b) A female's paracallosal sex skin, with oestrous vesicles around the edge.

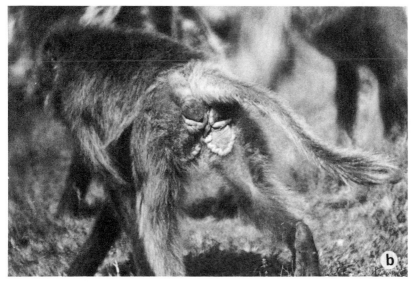

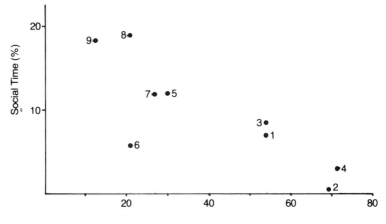

Figure 10   Time spent in social interaction, plotted against time spent resting, for 9 species of Old and New World Monkeys. Sources of data: (1) *A. geofroyi* and (2) *A. paliatta,* Richard (1970); (3) *C. badius,* Clutton-Brock (1972); (4) *C. guereza,* Dunbar and Dunbar (1974a); (5) *P. anubis* (mean of 4 populations), (6) *P. cynocephalus* (mean of 3 populations), (7) *P. ursinus* (mean of 2 populations), all from Dunbar and Sharman (unpublished); (8) *P. papio,* Sharman (1981); (9) *T. gelada* (mean of 3 populations), Iwamoto and Dunbar (1983).

These strategies require anatomical and physiological specializations. They also impose time budget constraints: rumination, in particular, requires time that cannot be devoted to any other activity. Impala, for example, spend about 15% of every 24 hours ruminating (Jarman and Jarman 1973); klipspringer spend about 10% and feral goats about 12% (unpublished data). Those primate species that pursue a ruminant-like strategy by carrying out a great deal of foregut fermentation seem to spend more time resting and less time in social activity than do congeneric species that are more frugivorous. *Colobus guereza,* for example, spend more time feeding and resting and less time in social activity than the more frugivorous *C. badius* (compare Dunbar and Dunbar [1974b] with Clutton-Brook [1972]), while *Hylobates syndactylus* spend significantly more time both feeding and resting than the more frugivorous *H. lar* (Raemakers 1979). Time committed to resting for dietary reasons is time that cannot be devoted to social interactions. Indeed, time spent in friendly social interaction does seem to be negatively related to time spent resting among the Old and New World monkeys (Fig. 10: $r_s = -0.744$, $n = 9$, $p = 0.022$).

An alternative solution to the problem of digesting foliage is simply to accept the fact of a low nutrient gain per unit weight of food ingested

Table 3

Activity budgets of the Sankaber and Bole gelada (based on scan samples taken at 20-min intervals), with comparable data for *Papio anubis* (based on data from the Bole study area).

| Activity | Time spent in each activity (%) | | Anubis (Bole) |
| | Gelada | | |
	Sankaber	Bole	
Resting	13.8	26.3	29.7
Social	20.5	18.5	14.3
Moving	20.4	17.4	27.1
Feeding	45.2	35.7	28.0
Number of records	21182	3358	2546

and to compensate by eating more of it. Bulk feeding is practiced by horses and other non-ruminant ungulates (Janis 1976) and by geese, one of the few grazing birds (Sibley 1981). It is a strategy that is probably quite easy to evolve and is the solution that the theropithecines opted for.

Behavioral Consequences of Graminivory

The gelada's diet has important consequences for a number of ecological variables (see Iwamoto and Dunbar 1983). I here consider three (activity budgets, day journey, and population density) that are likely to place particularly important constraints on the animals' reproductive strategies.

ACTIVITY BUDGET

Table 3 gives a breakdown of the activity budgets of the Sankaber and Bole gelada. Approximately 45% of their time is spent feeding, with an additional 20% spent moving from one feeding site to another. Thus, nearly two-thirds of the animals' time is devoted to obtaining the nutrients they require each day. This is rather more than is required by the more frugivorous *Papio* baboons (see final column of Table 3) and is a direct consequence of the gelada's strategy of bulk feeding on a relatively poor quality food source.

Gelada do not occur below an altitude of 1500 m asl: the main study area at Sankaber, for example, lay at an altitude of 3300 m. Ambient temperatures are consequently low (commonly below freezing at night) and thermoregulation is an important problem for the animals. The amount of time spent feeding increases with altitude. This is a consequence of the fact that metabolic costs of maintaining body temperature increase as the ambient temperature declines with increasing altitude. The problem is compounded by the fact that forage quality declines as altitude increases, so that further time has to be devoted to feeding to offset the shortfall in the nutritional intake.

Additional feeding time is taken out of resting time in order to conserve social time (see Dunbar and Sharman 1983). Once resting time is used up, however, any additional feeding time would have to be taken out of social time. This would have serious consequences for the integration of the social groups. In practice, it is unlikely that gelada live at altitudes very much higher than those at Gich, so that the erosion of social time may not become a serious problem. However, even at the altitudes of Sankaber and Gich, the animals have relatively little leeway for budgeting their social activity. At Gich, for example, they spend only about 5% of the daytime resting, and thus have at most this proportion of time by which to increase their social activity. Since animals probably need some time during the day free of all activity for rest, social time at Gich may already be at its maximum.

DAY JOURNEY

Day journey length is a linear function of band size in the three study areas, indicating that an animal has to cover a more or less constant area each day to obtain its daily nutritional requirements. This relationship is known to occur in other species of primates (see Waser 1977, Sharman and Dunbar 1982). In the case of the gelada, this partly results from the fact that the additional energy requirements due to altitude are fortuitously almost exactly offset in our sample of habitats by an altitudinal cline in grass density such that, as altitude increases, the animals have to search proportionately less far afield to find the same quantity of grass to harvest. Thus, metabolic, ecological, and demographic factors interact to influence the distance that animals must cover during the day. In general, however, the bigger the group, the longer the day journey.

The longer the day journey, the greater the time that has to be devoted to moving, which will, in turn, place further constraints on the animals' time budget. The only way in which the time budget can be adjusted while conserving social time is to reduce the band size in order to reduce the distance that has to be covered each day. There is some evidence to suggest that bands at Gich are smaller than those at the lower altitude of Sankaber (see Table 2).

An important consequence of this decrease in band size will be that small sample biases will become more significant. These will affect a number of demographic variables (birth rate, neonatal sex ratio, adult mortality rates) that are directly relevant to the reproductive strategies of both sexes.

POPULATION DENSITY AND RANGING PATTERNS

Gelada live at very high densities. The mean density of animals in a band's ranging area was 72.5 animals/km^2 ($n = 7$ bands), the variance around this being relatively small (s.d. $= 11.62$). Ranging areas are linearly related to band size. Because range areas overlap extensively (in some habitats, such as Sankaber, no areas were used exclusively by any one band), the number of different animals using a given area may be very much higher. This high stocking rate is largely due to the gelada's specialization as a grazer, since grass is the main component of the primary vegetation biomass throughout the species' range. It is also an indication of how efficiently gelada can utilize grass. They are so well adapted to their habitat that they are able to maintain densities that are 3–4 times greater than those of the more frugivorous *Papio anubis* in areas where the two species are sympatric (Dunbar and Dunbar 1974b). In fact, gelada are invariably the dominant component of the non-domestic faunal biomass throughout the Ethiopian highlands: they are able to maintain significantly higher densities than any of the antelope species with which they are sympatric (Dunbar 1978a, 1984a).

Because grass is a more or less uniformly distributed resource, gelada gain little by defending territories. The high population densities, combined with day journeys that are long relative to the small size of their home ranges, result in an extensive overlap in ranging areas and a consequent frequent mixing of units from neighboring bands. Range overlap is less, and the degree of intermingling of units from different bands correspondingly lower, in areas where band sizes are small and day journeys consequently shorter. At Sankaber, where densities and band sizes were at their highest, range overlap and intermingling were at their greatest and individuals probably had considerable knowledge of the demographic state of a large proportion of the bands in the local population. This knowledge, as we shall see, may be of some importance to males in their choice of reproductive strategy.

4 Demographic Processes

This chapter summarizes the main demographic processes and life-history variables and provides the demographic information on which most of the analyses of reproductive strategies depend. The results presented here are based on analyses given by Dunbar and Dunbar (1975), Ohsawa (1979), Dunbar (1980a), and Ohsawa and Dunbar (1984). These papers should be consulted for details of the deduction of causal inferences.

Life-History Variables

SURVIVORSHIP AND MORTALITY

Figure 11 shows estimated survivorship curves for males and females of the Sankaber population. The curves follow the pattern typical of most primate species: mortality among immatures is highest among infants and plateaus out during the juvenile period before increasing progressively with age during adulthood. Among adults, the mortality rate is higher among males than it is among females. The life-expectancy at birth is about 12.3 years for males and 13.8 years for females.

It is worth pointing out that the neonatal death rate is exceptionally low in this population: survivorship was estimated to be 91% to 18 months of age and 88% to 4 years of age. In contrast, survivorship to 4 years was only about 20% in *Macaca sinica* (Dittus 1975) and 45% in *Papio cynocephalus* (Altmann 1980). Even provisioned populations of macaques fare rather poorly by comparison: survivorship to 4 years of age was 79% for both the Caribbean *Macaca mulatta* population (Sade et al. 1977) and the Takasakiyama *M. fuscata* population (Masui et al. 1975).

Infant mortality at Sankaber correlated with the number of wet season months that infants had to undergo in their natal coats. Adult mortality was also highly seasonal, being greatest in wet season months and least in dry season months (Dunbar 1980a). The main cause of mortality was probably respiratory and other infections contracted during the climatically adverse wet season when the animals were subjected to constant soakings at temperatures below 0°C. The other main cause of mortality was infestations by the tapeworm *Multiceps servalis*. Similar

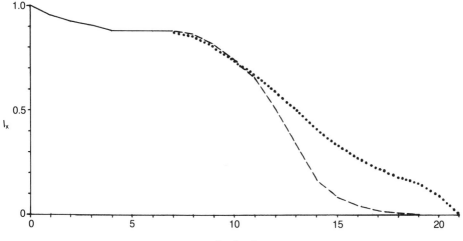

Figure 11   Estimated survivorship (l_x) of male and female gelada. The sexes are not distinguished prior to puberty; thereafter, the dashed line indicates males, the dotted line females. Based on mortality rates at Sankaber given by Dunbar (1980a, Table 10).

conclusions were reached by Ohsawa (1979) in studies of the Gich population. The importance of climate as a determinant of mortality was confirmed by Ohsawa and Dunbar (1984), who found that mortality among both immatures and females was higher at Gich than at the lower altitude of Sankaber where the climate was less extreme. Male mortality rates, however, did not correlate with altitude; rather, they appeared to correlate with the rates of harem takeover, suggesting perhaps that the physiological stresses imposed on males by these social changes might reduce the life expectancy of the older animals. There was no evidence to suggest that food shortage was ever a cause of mortality: indeed, Iwamoto (1979) found that the Gich population was in energy surplus throughout the year.

Fecundity

The gelada have a surprisingly low birth rate compared with other baboons and macaques. Females give birth for the first time at about 4 years of age and, at Sankaber, continue to produce offspring thereafter at intervals of approximately 2.14 years. Birth rates at Sankaber were found to be cyclic, with a cycle length of approximately 4 years (Fig. 12). The variability in the birth rate from year to year was attributed

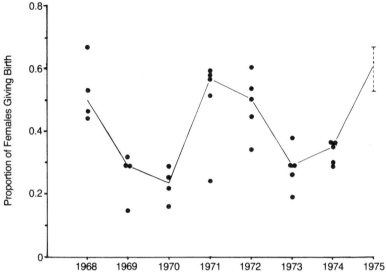

Figure 12   Birth rate per female in different years for 5 Sankaber bands. A minimum-maximum estimate is given in 1975–76 for the Main band based on the known and maximum likely number of pregnancies, repectively. (Reproduced from Dunbar 1980a, Fig. 3).

largely to the effect of climatic variables, since there was a negative correlation between birth rate and rainfall in any given year (Dunbar 1980a). The data suggest that increasing climatic severity at such high altitudes results in high rates of abortion due to cold stress. Evidence to support this conclusion came from an analysis of the timing of conceptions in relation to monthly rainfall (Dunbar 1980a, Fig. 5). This revealed that, within any given year, the peak in conceptions coincided with the peak in rainfall, suggesting that the animals' improved condition as a result of the vegetation growth at the onset of the rains triggered ovulation. (The availability of high quality food is known to trigger ovulation in a number of species: see Sadleir 1969, Belonje and van Niekerk 1975.) Further evidence to support the hypothesis comes from a comparison of the populations at Bole, Sankaber, and Gich (Ohsawa and Dunbar 1984). Similar results have been reported for coypu (Gosling 1981).

Births are more highly synchronized within reproductive units than they are between units of the same (or different) band(s). While the rather generalized level of synchrony within bands as a whole can be attributed to the tendency for the onset of the rains to trigger concep-

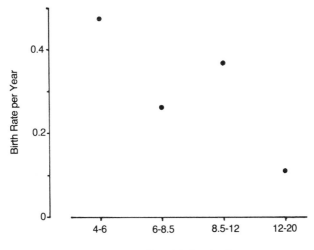

Figure 13   Age-specific birth rates. Based on data given in Dunbar (1980a, Table 11).

tions, the much higher degree of synchrony within individual harems of females must be attributed to social factors. Takeover of harems by males from all-male groups was identified as the primary cause of this synchrony: the inter-birth interval is significantly shorter than normal following the takeover of a unit by a new male (Dunbar 1980a, Fig. 6; see also Mori 1979a, Mori and Dunbar 1984).

The birth rates of individual females are also influenced by two other factors, namely, a female's age and her dominance rank within the unit. Fecundity (Fig. 13) follows the general pattern reported for primates, being highest in younger females and lowest in old age. Fertility is known to decline with age in a number of mammalian species (e.g. humans, Fédération CECOS et al. 1982; horses, Laing and Leech 1975). To a large extent, this is probably a consequence of increasing ovulatory irregularity with age (cf. Graham et al. 1979). The second factor was the effect of dominance rank on fecundity. Dunbar (1980b) found a negative correlation between the number of offspring a female had and her dominance rank. Detailed analyses showed that stress from repeated harassment by higher-ranking females caused ovulatory failure in low-ranking individuals; consequently, they required more oestrous cycles to conceive than did high-ranking females, whose inter-birth intervals were proportionately shorter. This relationship will be considered in more detail in Chapter 6.

Migration Rates

Most migration out of bands occurred as a result of the fission of bands when part of a band moved into a new home range. Bands at Sankaber and Gich underwent fission about once every 8.5 years on average (Ohsawa and Dunbar 1984). Actual gene flow between bands, however, occurs as a result of only two processes: the migration of individual reproductive units from one band to another and the transfer of individual animals (usually males) between bands. The two processes are of about equal importance, together accounting for an average of 0.016 migrations per individual per year. On the basis of the genetic similarity between neighboring bands, Shotake (1980) estimated gene flow between bands to be of the order of 5% per generation. The observed migration frequencies suggest a rate of gene exchange of the order of 6.4% per generation, which agrees reasonably well with Shotake's estimate from the distribution of protein polymorphisms.

Females were never observed to migrate between bands as individuals; indeed, they were only very rarely observed to move from one reproductive unit to another even within the same band (Dunbar 1980a, Ohsawa 1979). Females thus remained within their natal units together with their female relatives throughout their lives, whereas males first joined an all-male group as juveniles or subadults and then later moved back into a reproductive unit, sometimes in a different band.

Determinants of Demographic Structure

Ohsawa and Dunbar (1984) found that the birth rate determines the future adult sex ratio as a consequence of the difference in the ages at which the two sexes mature: in expanding populations, the adult sex ratio becomes more disparate in favor of females as the birth rate rises. This effect is exacerbated in the Simen populations by contrasting changes in the mortality rates of the two sexes that may themselves in part be a consequence of climatic differences between the habitats.

Mean harem size does not correlate consistently with any environmental or demographic variable. Harem size is not, for example, related to the adult sex ratio (Kendall's $\tau = 0.241$, $z = 1.350$, $p = 0.177$, $n = 16$ bands at Sankaber and Gich). This suggests that harem size may be a consequence of social factors that influence harem fission rates.

The tendency for females to remain in their natal units means that, except in cases of very high mortality, reproductive units will tend to increase in size with time. At Sankaber, the time taken for the female complement of a unit to double is equal to one-and-a-half times the inter-birth interval (2.143 years) plus the average age at first birth (4.0 years), compensated for survival to breeding age. This is roughly equivalent to the sum of the likelihoods that each successive infant will be female, multiplied by the time it takes to mature. With a sex ratio at birth of exactly 50:50 (Dunbar 1980a), a survivorship of 88% to the age of 4 years, and a mean mortality rate of 0.077 per female per year, the doubling time is 6.73 years at Sankaber (less at Bole, more at Gich due to shorter and longer inter-birth intervals, respectively). This is equivalent to an annual compound rate of increase of 10.9%. The average female could thus expect to see her harem increase in size by a factor of nearly 2.5 during her breeding life. (Expectation of life at reproductive maturity is $e_4 = 9.8$ years). With a mean harem size of about 4 females, the average female would die in a unit of about 10 females, and many would be in very much larger units. Since only 15% of units had more than 6 females (Fig. 5) and the largest unit ever observed in any population contained only 12 females, it is clear that units must undergo fission from time to time.

At Sankaber, units were observed to undergo fission at the rate of 0.120/unit/year (or at a rate of 0.156/unit/year for units with 4 or more females). This suggests that each large unit (≥4 females) would expect to undergo fission once in every $1/0.156 = 6.4$ years on average (or approximately once during the time it takes to double its size). If the distribution of harem size is determined largely by the rate of fission, we should be able to predict the distribution of large units observed in the population by using the observed growth rate for units (10.9% *per annum*) and the observed fission rate (0.156 per unit *per annum*) to determine a survivorship function for a unit that starts with 4 females. From this, we can calculate a size-specific probability distribution. This function is given in Table 4, together with the expected distribution of harem sizes generated by it and the actual observed distribution found in the main study bands at Sankaber. The two distributions are remarkably similar except in their tails: however, they do not differ significantly ($\chi_4^2 = 8.496$, $p = 0.178$). This suggests that harem size is determined only by female birth and death rates and a statistically random overlying fission process. (Note that this does not mean that fissions occur for no reason, but rather that we cannot predict which units will undergo fission on the basis of size alone. The non-random causes of fission will be discussed in Chapter 11.) One consequence of this is that, other things being equal, harem size will tend to oscillate around a general mean, and so (in the long term) to remain within a specified range of variation.

Table 4

Frequency distribution of large units (≥4 reproductive females) in the main study bands at Sankaber (data for both studies pooled), compared with a distribution predicted by a random fission process (constant fission rate of 0.156/unit/year irrespective of size) at a population growth rate of 10.9% *per annum*.

Harem Size	Conditional Probability of Occurrence[a]	Frequency in Population Expected	Observed
4	0.323	12.0	12
5	0.186	6.9	8
6	0.116	4.3	5
7	0.079	2.9	5
8	0.041	1.5	4
9	0.031	1.1	1
10+	0.169	6.3	2
Total	1.000	37.0	37

[a] The probability that a unit starting with 4 females would survive to the indicated size without fission was determined, given the observed fission and growth rates for the sample population. The conditional probability of occurrence was then calculated by expressing the survivorship to a given size as a function of the sum of the survival probabilities for all harem sizes.

5 Social Structure of Reproductive Units

Gelada reproductive units are closed social microcosms (see Dunbar and Dunbar 1975, Mori 1979b). Consequently, an individual's social and reproductive strategies are constrained by the behavior of the other members of its unit. Understanding the principles that underlie social relationships between the adult members of a unit is thus critical for any analysis of the reproductive strategies of both males and females. In this chapter, I draw on analyses given by Dunbar (1979b, 1980b, 1982c, 1983b,c,d) and also present some new data.

Social Relationships among Females

Analysis of the structure of relationships between reproductive females shows that females tend to form dyads whose members spend most of their time grooming each other (Dunbar and Dunbar 1975). The network of relationships may extend to three, occasionally four, females, but the mean size of these clusters is two individuals. A sociogram of a typical reproductive unit is given in Figure 14 and clearly illustrates this tendency for the females to form grooming pairs.

Detailed analysis of the relative ages of the females in these grooming dyads suggests that they consist mainly of mothers and daughters since they consist almost exclusively of an older female and a younger one (Dunbar 1979b). This inference was later confirmed by an analysis of grooming patterns in a long-established captive group in which all mother-daughter relationships were known (Dunbar 1982c). In general, females showed a marked preference for grooming with their mature daughters. A comparison of the observed distribution of the sizes of grooming networks with that predicted by the theoretical distribution of matriline sizes (given the observed life-history characteristics of the population) indicates that females rarely interact with individuals who are not members of their immediate matriline (Dunbar 1983b). Demographic analyses suggest that the average matriline size should be just two mature females, a size that corresponds exactly to the observed mean number of females in a grooming cluster.

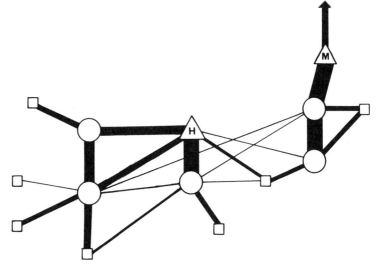

Figure 14   Sociogram of the 1974–75 unit N12, a typical gelada reproductive unit, showing the relative frequencies with which the adult and subadult members of the group interacted. Frequencies of interaction are proportional to the width of the line joining any two individuals and were estimated from 304 scan samples of the unit during social periods. Triangles indicate mature males, circles indicate reproductive females and squares indicate immature offspring. H = harem male, M = subadult natal male.

An important corollary of this is that females who do not have living female relatives do not interact socially with the other females in their unit to any significant extent. Instead, they tend to be socially peripheral, grooming mainly with their immature offspring. This conclusion was confirmed by observations of the social interactions of two females following the deaths of their female grooming partners: in neither case did they attempt to develop new grooming relationships with the other females of their units.

At least part of the reason why a female's circle of interactees is so restricted is that the time she has available to interact with other females is limited (see also Mori 1979c). At Sankabar, females spent an average of 19.9% of their day in social activity. Although in theory a female could spend an additional 8.6% of her time enlarging her social circle instead of resting, she can only do this if (a) she is motivated to do so and (b) other individuals are willing to interact with her. In fact, social time is conserved across habitats among baboons in general (Dunbar and Sharman 1983), suggesting that gelada do all the social activity they can do.

Of the female's potential social time (approximately 35% of her whole day, of which about half is actually spent interacting), an average of 11% is spent interacting with the unit's male, 14% with her own immature offspring and 27% with her female relatives. On average, a female spends 21% of her potential social time interacting with her main female grooming partner (i.e. the female she grooms with most), leaving only 6% of her time to devote to other female interactees.

Social networks in gelada units are thus small and confined to very closely related females. Other reasons why females limit their social networks so severely will be discussed in Chapter 7.

Dominance Relationships among Females

Approach-retreat agonistic encounters in non-social situations were used to rank the reproductive females of each unit on the basis of wins and losses. These hierarchies turned out to be strictly linear with no cases of non-transitive dominance relationships and surprisingly few cases of wins against the hierarchy. (A *non-transitive* dominance relationship is one in which A dominates B and B dominates C, but C is able to dominate A; *wins against the hierarchy* are cases in which the subordinate animal is able to defeat an individual that normally dominates it.)

The stability of these hierarchies appears to decline as harem size increases. Figure 15 shows that the proportion of unstable dyads (i.e. those in which wins against the hierarchy are observed) increases with harem size ($r_s = 0.671$, $n = 11$; $t = 2.712$, $p = 0.024$ 2-tailed). Because there is no correlation between the number of encounters sampled per unit and the proportion of wins against the hierarchy ($r_s = 0.019$, $n = 11$, $p = 0.886$ 2-tailed), nor one between harem size and the number of encounters sampled ($r_s = 0.506$, $n = 11$, $p = 0.112$ 2-tailed), it is clear that harem size is the key variable. I interpret this relationship as being due to the decreasing familiarity between members of large units; their status relative to each other is thus likely to be more uncertain on the occasions when they do interact, and is therefore more likely to be challenged. In contrast, members of small units interact more often and have little need to keep testing their relationships, as these are mutually well known and hence more predictable.

Before accepting this explanation, however, we need to dispose of at least three alternative hypotheses which can be suggested to account for the occurrence of wins against the hierarchy. These are: (1) the outcome of a particular encounter depends on the desirability of a given resource to the normally dominant member of the dyad, thus determining whether or not she will let the subordinate animal win on that

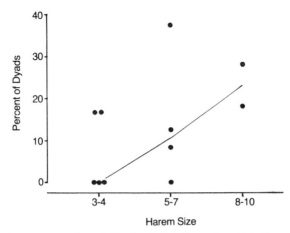

Figure 15 Percentage of unstable dyads (i.e. those in which there were "wins against the hierarchy"), plotted against harem size for 11 units whose dominance hierarchies were determined during the 1974–75 field study.

occasion (see Parker 1974, Popp and DeVore 1980); (2) dominance relationships between adjacently ranked animals are the most likely to be unstable if they are of similar physical size (as is the case in many caprids: see Geist 1971, Schaller 1977, Dunbar and Dunbar 1981) and, because there are more adjacently ranked animals in larger groups, a higher proportion of wins against the hierarchy will result; and (3) since kin selection could result in a female behaving altruistically by tolerating subordination to close relatives while challenging less closely related females more often (see Datta 1981), the observed distribution could be a consequence of the fact that females in large groups are, on average, less closely related to each other than are females in small groups.

The first of these hypotheses may easily be dismissed because, although the encounters occurred in non-social (i.e. mainly feeding) contexts, relatively few of them were actually contests for access to food resources: most of them seemed to involve sheer harassment (see Dunbar 1980b). This conclusion is supported by observations of a captive group during feeding sessions where, depite a high rate of encounters, very few interactions resulted in the winner taking food from (or even occupying the feeding site of) the loser, even when the loser moved to a new site (see Dunbar 1982c). Moreover, if the hypothesis were true, there would be no reason to expect the rate of wins against the hierarchy to increase with group size (if only because the variance in group

Table 5

Frequency distribution of the number of dominance ranks separating members of an unstable dominance dyad, compared to an expected distribution based on the assumption that unstable dyads occur at random in proportion to the availability of the various rank differences in the sample population (calculated by enumeration).

	Number of ranks separating members of dyad								Total
Number of Dyads	1	2	3	4	5	6	7	8	
Observed	5	7	3	1	1	1	—	—	18
Expected	5.6	4.4	3.2	2.0	(	2.8		)	

size is trivial by comparison with the sizes of the herds which the reproductive units form when foraging).

The second hypothesis can be tested by comparing the distribution of unstable dyads with an expected distribution based on the assumption that such dyads occur at random among the pairings available in the sample units (see Table 5). It is clear that unstable dyads are distributed at random ($\chi_4^2 = 2.288$, $p = 0.680$). This result is more consistent with the familiarity hypothesis, since individuals that groom regularly tend to rank next to each other in the dominance hierarchy (see below).

Finally, kin selection can be ruled out on the grounds that wins against the hierarchy were not significantly less common between females who groomed each other regularly than between those who did not. In the sample of 11 units, there were 4 unstable dyads among 15 grooming dyads (26.7%), but only 12 in 91 "non-grooming" dyads (13.2%). The difference is not statistically significant (comparison of two binomial probabilities, $p = 0.183$), although this could reflect the small size of the sample for grooming dyads. Nonetheless, since the distribution is in the opposite direction to that predicted by an explanation based on kin selection, we can safely rule out any such hypothesis. (Note that familiarity in the sense posited here need not have anything to do with the frequency of grooming; rather, familiarity is a function of exposure to another individual's behavior and capabilities and thus has more to do with the perceived predictability of behavior.)

The implications of this are twofold. First, there is an element of inertia in the dominance hierarchies of female gelada, especially in small units, that tends to maintain the status quo. Secondly, the amount of effort required to maintain rank is relatively low, certainly much less than that required in a more competitive situation. This is commensurate with the low rate of encounters among females (mean of only 0.170 encounters per dyad per hour for 53 dyads in 7 units, of which only 5.5% are involved physical contact). On average, each female initiated 3.48 encounters per day, of which only 0.19 (one every five days) escalated to include biting or a prolonged exchange of visual and vocal threats. The costs of maintaining rank are clearly very small. These results are consistent with Rowell's (1966) finding for captive *Papio anubis* that the hierarchy tended to be maintained by the behavior of the subordinate rather than the dominant animals. Mori (1977) found an analogous increase in the proportion of unstable dyads as the number of females increased in the Takasakiyama troop of *Macaca fuscata*. Similarly, Guhl (1968) noted that, among chickens, social stability reduced both the amount of fighting and the frequency of rank changes, apparently because the members of a group that remained together became increasingly familiar with one another.

The final point to be considered is the important question of what determines a female's rank in the hierarchy. Detailed analyses have shown that a female's rank in her unit depends on a three-step rule (see Dunbar 1980b):

(1) A young adult female occupies a rank that is dependent solely on her aggressiveness with respect to other young adult females;

(2) A female of any other age class occupies a rank that is immediately below that of her highest ranking female relative;

(3) A female of one of these other age classes who does not have a mature female relative in the unit will occupy a rank at the bottom of the hierarchy commensurate with her own intrinsic aggressiveness vis-à-vis prime age (i.e. young adult) females and other females who have no relatives.

Thus, entire matrilines rank intact with respect to each other because their relative ranks are dependent solely on the aggressiveness of their highest-ranking members. Matriline rank is not a consequence of matriline size; indeed, in two units, the dominant female remained dominant after the death of her matrilineal relative even though she then had no female relatives, whereas the next ranking female did. Rank *within* a matriline depends on an individual's aggressiveness, and this is an inverted-J-shaped function of age (Fig. 16). (Intrinsic aggressiveness was measured by the frequency with which a female initiated agonistic encounters with individuals in other reproductive units, since this was less likely to be confounded by dominance relationships among the females: see Dunbar 1980b.) The females in each age class were divided into two groups: those that had a female partner with whom

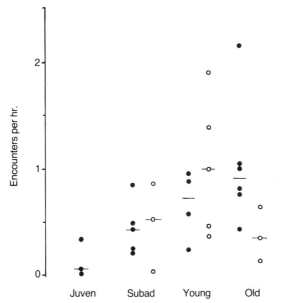

Figure 16 Rates at which females of different age classes initiated agonistic encounters with neighboring reproductive units. Females of each age class are distinguished according to whether they were members of a grooming dyad (filled circles) or not (open circles). Source of data: 11 reproductive units sampled in 1974–75.

they spent at least 10% of their potential social time grooming and those that did not. It is clear that the presence of a grooming partner (i.e. a close matrilineal relative) significantly increases the intrinsic aggressiveness of old females, though it does not affect that of females of other age classes. In other words, old females are more willing to initiate agonistic encounters if they have a female relative on whom they can count for support; without such a relative, the female's willingness to challenge others (and hence her potential dominance rank in the unit) is comparable to that of subadult females.

Neither a female's dominance rank nor her aggressiveness are in any way influenced by the extent to which she interacts with the unit's male (see below).

Social Relations between the Male and His Females

On average, a harem male spends 53% of his potential social time interacting with his females and only 2% interacting with other members of his unit (mainly juveniles). He does not interact equally with all of his females, but tends to spend most of his time grooming with one particular female. My main concern here is to determine which female becomes the male's main grooming partner and why. The question of why a male should spend most of his time grooming with only one of his females will be deferred to Chapter 10 where its relevance to male reproductive strategies is discussed in more detail.

Analysis of data from 11 reproductive units showed that the male's main partner was invariably one who had no female relatives with whom to groom (Dunbar 1983c). The situation is a little more complicated than this, in that the highest-ranking female does have priority of access to the male if she wishes to exercise it, though she usually does so only if she herself has no female relatives. Whenever the dominant female was *not* the male's main partner, then the most most dominant female without a female grooming partner usually was. This conclusion is supported by Kummer's (1975) studies of artificially convened groups: when some of the females deserted their male following the addition of a second male to the compound, the female that formed the strongest relationship with the new male was the most dominant of the deserters. The result was that the most dominant female in each of the groups monopolized the two males.

Generally speaking, the male's partner female puts about as much effort into maintaining their relationship as the male does. Other females show much less interest in him: they tend to respond rather perfunctorily to the male's social advances, initiating relatively few of their interactions with him, responding less frequently to his requests to be groomed, and terminating interactions at an early stage (Dunbar 1983c).

In fact, the relationship between the male and his female grooming partner closely resembles the relationship found between females in grooming dyads, and there was much to suggest that, socially at least, the male's partner treats him in much the same way as she would treat any female grooming partner. There was a correlation between the male's age and that of his female partner, suggesting that the relationship may be maintained over a considerable period of time. Indeed, in all 7 observed cases of harem takeover, the former harem-holder continued to groom predominantly with his previous partner female after his displacement as the unit's breeding male (in one case for more than two years). These findings are best interpreted in terms of the female's preference for maintaining her existing social relationships until a more desirable social partner appears (see Chapter 8). Such partners can only be daughters (recall the female's general preference for interacting with mature female relatives). Clearly, a female who gives birth to a succession of male infants at two-year intervals will remain for substantial periods without any incentive to break off her relationship with the male.

Although the unit's dominant female exercises her prerogative of social access to the male only occasionally, she invariably maintains strict control over access to him during non-social periods by actively interposing herself between him and the other females. Consequently, even the male's partner female has to run the gauntlet of more dominant females if she wants to gain access to the male outside of social periods. We shall see in the following chapters that this has important consequences for the reproductive behavior of the females.

A further indication of the relative value that a female attaches to the various members of her unit is given by the frequencies with which she monitors them. Although females in general tend to monitor the male more closely than they do any of the females in their unit, they show clear preferences among the females for their main grooming partner and the dominant female. The prominence of the male and the dominant female in this respect reflects their role in controlling progressions by the unit (see Dunbar 1983d). A female's attention thus seems to be torn between maintaining contact with the unit's central decision-makers and ensuring that she does not become separated from her preferred social partner. It is important for a female to maintain proximity to an alliance partner because that individual is the one most likely to come to her aid when she is threatened by members of another unit (Dunbar 1980b; see also Chapter 7). Contact is also maintained vocally through the use of a complex repertoire of contact calls (see Richman 1978, Kawai 1979b). These calls, which exhibit some of the acoustic properties of human speech (Richman 1976), are used to maintain and service relationships between social partners in a way analogous to human coversation (unpublished data).

Males become increasingly less diligent in their attempts to groom with all their females as they get older (Dunbar 1983c; see also Fedigan 1972). This is probably because younger males, who have only recently acquired their harems, are under some pressure to establish relationships with their females and therefore try to groom with more of them regularly. Older males, who have already established these "minimal relationships," seem to be less inclined to pursue interactions with females that are generally unwilling to interact with them.

The male neither supports nor herds his females in a way that is consistent with their dominance ranks, indicating that he does not influence the females' ranks within the unit. In fact, he tends to support the lowest-ranking females most often (Dunbar 1980b; see also Bramblett 1970) and to herd the highest-ranking females most (Dunbar 1983c). Although the male is, without exception, the most dominant member of his unit, he by no means has everything his own way. Dunbar and Dunbar (1975), Kummer (1975), and Mori (1979b) have all reported that high-ranking females often disrupt social interactions between the male and subordinate females. Furthermore, although females who stray from their units are herded by the male, such behavior usually results in some or all of the females chasing the male, often over a considerable distance.

Sexual relationships between the male and the females are similarly low-key (Dunbar 1978b). Oestrus affects neither a female's dominance rank nor the general pattern of her social interactions. The male does not groom with a female more when she is in oestrus than at any other time. Indeed, grooming rarely follows copulation, except between the male and his main social partner. Of a sample of 71 copulations involving 14 cycling females, only 22 (31%) were followed by grooming (mostly by the female). This is about what would be expected if grooming only involved partner females (expected $= 28.6\%$; $\chi^2_1 = 0.199$, $p > 0.60$).

Stability of Social Relationships

At least in the short term, the structure of relationships within gelada reproductive units remains remarkably stable. Comparisons of these relationships before and after a variety of potentially disruptive events (including the presence of an oestrous female, births, deaths, and takeover by a new male) indicated that their patterns remain largely unchanged (Dunbar 1979b). Similarly, comparison of the structure of relationships over periods of 5–8 months revealed no marked differences. How long such relationships are likely to remain stable is unclear, but there is evidence to suggest that even over periods of 2–3 years the structure of social relationships remains much the same, despite considerable demographic changes in the composition of the unit (see Dunbar 1979b).

This stability is in part a consequence of the fact that females will intervene in interactions between their main social partners and other members of the unit, dominance rank here being the key factor. Displacements in social contexts were in the direction predicted by dominance ranks based on data from non-social contexts in 27 out of 30 instances: such a distribution would be significantly unlikely to occur by chance if social encounters were independent of dominance ranks (binomial test, $p \ll 0.001$). High rank thus gives a female the opportunity to monopolize whichever individual she prefers (see also Spivak 1971, Bramblett 1970), thereby contributing to the stability of these relationships over time.

This is not to say that there are no important changes in the social structure of reproductive units as a direct consequence of demographic changes (to be discussed in Chapter 10). The basic principles that determine the general pattern of dyadic relationships both between females and between the male and his females remain the same despite a wide variety of significant demographic and reproductive changes. As a result, the *qualitative* pattern of dyadic relationships remains stable over time. We shall see in Chapter 10 that there are significant *quantitative* changes in the overall pattern of dyadic relationships within a harem as its size increases. To use Hinde's (1975, 1983) terminology, the surface structure of relationships changes in response to demographic changes, but the underlying deep structure (the rules determining an individual's social preferences) remains unchanged.

One consequence of this stability is that a unit's integrity over time is not dependent on the presence of the male (as is the case in *Papio hamadryas* baboons, for example: see Kummer 1968). Dunbar and Dunbar (1975) argued that it is the strong cohesive bonds between the females themselves that hold a unit together and predicted that units would remain intact even if the male was removed. This prediction was later confirmed experimentally at Gich (Mori 1979c). Mori also reported that another unit whose male disappeared (presumably died) remained intact until taken over by a new male some weeks later.

Summary

The social relationships of gelada females are dominated by their preference for interacting with close female relatives. Time budget constraints limit the number of adults with whom a female can interact. As a result, a strong tendency is evident for the females to fall into distinct social dyads, with relatively little interaction between dyads. Within this context, the male functions much like any female member of the unit: his access to individual females is constrained by the female's own social preferences. Dominance relationships among the females of the unit are matrilineally determined, the relative ranks of the matrilines being determined by the relative aggressiveness of each matriline's most dominant member. The male is invariably dominant to all his females, but he does not influence their dominance ranks or social relationships. The importance of these relationships to females is reflected in their tendencies both to monitor their partners' movements closely so as to remain near them and to support their partners when they become involved in altercations with members of neighboring units. The females' tendencies to remain together in small groups probably serves to minimize interference from the many animals in gelada herds (in the sense suggested by Wrangham [1981], though in this case the unit *qua* coalition probably buffers the female against excessive harassment rather than providing her with access to some limited resource as suggested by Wrangham).

6

Constraints on Female Reproduction

In this chapter, I consider some important reproductive consequences of the social relationships among females that were described in Chapter 5. These consequences provide the reproductive context within which females make their strategic decisions. It is thus in the constraints imposed by these consequences that explanations for the behavior of the females are likely to be found.

Oestrus and Sexual Behavior

As far as we know, gelada do not differ significantly from other Old World monkeys in the general features of female reproductive physiology (though it is conceivable that there are quantitative differences in hormone titres). They do, however, exhibit a number of unique anatomical features. Unlike other baboons, female gelada do not undergo a cyclical swelling of the perineal areas of sexual skin. Instead, physiological oestrus (i.e. the imminence of ovulation) is indicated by the appearance of a chain of small, pink, fluid-filled beads or vesicles around the edge of each area of sexual skin, one of which is situated on the chest (Alvarez 1973, Dunbar and Dunbar 1974c). The only exception occurs at puberty: juveniles show a conspicuous swelling of the chest patch, together with a marked puffiness of the paracallosal sex skin (see Dunbar 1977b). There is no change in the color of the sex skins in relation to sexual state, though the paracallosal skin turns brilliant red during late pregnancy and the chest patch a bright red during lactation.

Mating occurs throughout the female's oestrous cycle, though its frequency increases markedly during the period around ovulation when the vesicles surrounding the sex skins are most fully developed. Copulations occur on average about once every 110 mins during peak oestrus, with 75% of these being initiated and actively solicited by the female (Dunbar 1978b). This contrasts with the situation reported for baboons and macaques, where the male usually initiates sexual interactions and commonly does so at a very much higher rate. Hausfater

(1975), for example, found that *Papio cynocephalus* copulate approximately once every 50 mins. It is easy to see that males living in one-male groups do not need to mate as frequently as animals in multimale groups where the risk of being displaced by a stronger male is very much greater. Once the gelada male has acquired his harem, he has uncontested sexual access to the females when they come into oestrus. Thus, he can afford to sit back and allow females to notify him when they are receptive. Nor does he need to fill the female's reproductive tract with sperm in order to maximize his chances of fertilizing her. This lack of competition for direct access to cycling females is reflected in the small size of the gelada male's reproductive organs relative to those of the average *Papio* male, in the small testis-to-body-weight ratio (Harcourt et al. 1981), in the general absence of sexual consortships, and in the fact that the male and female rarely groom each other following copulation. For the gelada, sex is perfunctory and interferes rather little with the participants' established pattern of relationships.

One marked quantitative difference in the males' sexual behavior was found, however: males are significantly less responsive to the sexual approaches of juvenile females than to those of older females, perhaps because juvenile females may, in at least some cases, be their own offspring (see Dunbar and Dunbar 1975). In addition, it is likely that, as with macaques (Rowell 1972), a female's first few oestrous cycles are infertile; mating with such females is likely to be of less value to the male in terms of reproductive output.

Reproductive Consequences of Dominance Rank

A female's dominance rank determines her reproductive rate at any given moment. This conclusion was based on a census of the numbers of living offspring less than 4.25 years old that females of different dominance ranks had in 1975. The relationship between dominance rank and number of offspring was checked in detail, both by excluding alternative hypotheses and more directly by using the relationship to predict the numbers of offspring in the units of the 1971–72 study bands at Sankaber. Since these tests were both exhaustive and stringent, there is little reason to doubt the validity of this relationship.

This relationship was used to determine rank-specific birth rates in order to assess the reproductive consequences of different lifetime rank trajectories (see Dunbar 1980b). There were, however, a number of problems with this particular analysis. One is that the estimate of rank-specific birth rates is based on an extended interval (4.25 years, nearly

Table 6
Mean birth rate and mean rank of females for 5 Sankabar bands

Band	Number of units	Mean harem size	Mean female rank	Birth Rate/Female/Year Observed	Corrected[a]
Main 1971	25	3.48	2.74	0.563	0.471
Abyss 1971	6	4.67	3.00	0.571	0.477
High Hill 1971	12	4.58	3.27	0.509	0.428
Main 1974	17	5.06	3.35	0.384	0.478
E1 1974	5	6.00	4.00	0.333	0.414

[a] Standardized to an average year to remove fluctuations in basal birth rate due to variations in climate (see Fig. 12).

half the average female reproductive lifespan). It is thus a weighted average of all the ranks occupied by the female during the sample period, whereas her dominance rank is taken to be that which she held at the end of the period. Since a female's rank is likely to undergo considerable change over her lifetime as the result of a variety of demographic events within her unit (Dunbar 1980b, Table 8; see also Gouzoules et al. 1982), the slope of the expression for rank-specific birth rates so obtained will be shallower than is in reality the case. In addition, the estimates will reflect year-to-year variations in the basal birth rate (see Fig. 12) and will not include any infants that died prior to the census. Both factors will tend to reduce the slope of the relationship.

Unfortunately, the number of units for which female dominance ranks are known is too small to yield unbiased estimates of actual birth rates due to the long inter-birth interval (approximately 2.1 years). An alternative approach is to analyze birth rates for entire bands in relation to the size of their harems: bands with more large units will suffer proportionately more reproductive suppression than bands with a larger number of small units. We can use these data both as a further test of the original hypothesis and as a way of obtaining a better estimate of the relationship between dominance rank and birth rate.

Table 6 gives the mean harem size, mean female rank, and mean birth rate per reproductive female for 5 Sankaber bands. Birth rates are also standardized to the same rainfall base (the average year) using the data in Figure 12 to provide a correction factor. A least squares linear regression of corrected birth rate on mean female rank gives:
$$y = 0.6093 - 0.0476x,$$

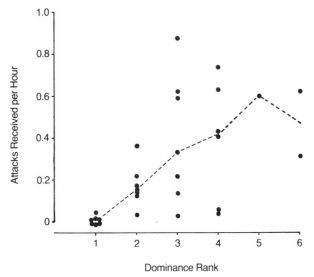

Figure 17   Frequencies per hour with which females of different dominance ranks were threatened by the other females in their units. The broken line joins the median values for each rank. Source of data: 7 reproductive units sampled in 1974–75.

with $r^2 = 0.551$ ($t_3 = -1.919, p = 0.075$ 1-tailed). The 95% confidence limits on the slope are -0.1265 and $+0.0313$, not too bad considering the small size of the sample but clearly rather closer to zero than one might wish for.

An investigation of the cause of this relationship revealed that the poorer reproductive performance of lower-ranking females was almost certainly due to an increased frequency of anovulatory oestrous cycles brought on by stress as a result of harassment by more dominant females (Dunbar 1980b). The physiology of this effect has been experimentally studied in talapoin monkeys by Bowman et al. (1978).

This explanation assumes both that the likelihood of ovulatory failure is a linear function of the frequency of threats received and that the frequency of threats received increases monotonically with declining rank. As far as the gelada are concerned, at least the second assumption seems to be justified. Figure 17 shows that the number of threats received is a cumulative function of declining rank (Spearman $r_s = 0.720, n = 30, t = 5.490, p<0.001$).

In summary, a female's birth rate declines with declining dominance rank, and it does so at a rate of approximately 0.048 births per year for each unit decrease in rank. (Bear in mind that the average birth rate at Sankaber is only about 0.45 births per year.) While this rate of loss is too small to have much effect on the reproductive outputs of females who rank near to each other in the hierarchy, the effect becomes increasingly severe on the lowest-ranking females as harem size increases. In a unit of 10 females, for example, the first- and fourth-ranking females can expect reproductive rates of the order of 0.562 and 0.419 births per year, whereas the tenth-ranking female can expect a rate of only 0.133 births. Translated into lifetime reproductive output, this would clearly represent a very considerable selective differential, and we may presume that low-ranking females would find it intolerable.

However, as with many findings of large differentials in the reproductive outputs of males (e.g. elephant seals, LeBeouf 1974; baboons, Hausfater 1975), this represents a static status-specific effect: it takes no account of the fact that an individual animal's rank alters over the course of its lifetime as a consequence of changes in its physical abilities and prowess (see, for example, Gibson and Guinness 1980). Other things being equal, an animal can expect to hold low rank early and late in life, achieving high rank for only a short period during its prime. Nonetheless, it is quite clear that the differential in instantaneous reproductive rate is so great that gelada females are likely to be under considerable selection pressure to find ways of minimizing the length of time for which they occupy low rank. The strategies that they pursue are examined in detail in the next three chapters.

Note added in proof

Data on observed birth rates for individual females in 11 Sankaber units in 1974–75 yield estimates of 0.444/year for females of ranks 1–2 ($n = 21$), 0.371 for ranks 3–4 ($n = 18$), and 0.251 for ranks 5–10 ($n = 16$). A least-squares linear regression gives a slope of birth rate on rank of $b = -0.032$. This does not differ significantly from the estimate obtained from Table 6 ($t_3 = 0.629$, $p = 0.716$).

7 The Female's Socio–Reproductive Strategies

In order to understand what the females are trying to do, we need to know what the null condition is around which their strategies of reproduction are built (i.e. the constraint-free strategy). In the present case, this is the state that would result if females relied only on their intrinsic natural fighting abilities. Other things being equal, a female would gradually rise in rank as she matured, reaching maximum rank during her prime (and in smaller units where few females are of similar age, this maximum will usually be rank 1), to decline again as she grew older. By and large, we would expect all females to pass through all ranks, occupying each one for approximately the same length of time as every other female (the vagaries of the demographic processes aside). Consequently, all females would have much the same mean rank over a lifetime and their reproductive outputs would be broadly similar.

By changing the rules, however, a female might be able to outcompete other females in her unit. To do so, she would need to step up her fertilization (or birth) rate, and this she could do in a variety of ways, some strategic, others purely tactical. In this and the following chapter, I analyze the female's strategic options, postponing discussion of her tactical options to Chapter 9.

In principle, a female has three main strategic options: (1) to form an alliance with another female in order to increase her dominance rank; (2) to become the male's main grooming partner, either in order to use him as an ally or to minimize the physiological stress incurred in having to bypass more dominant females to gain sexual access to him; and (3) to desert the male in favor of another male who has fewer females, thereby effectively increasing her dominance rank.

Coalition Formation

FUNCTIONAL IMPORTANCE OF COALITIONS

A female's dominance rank, as we have seen in Chapter 5, depends during at least part of her lifetime on the rank of either her mother or her daughter. Females gain this advantage through the protection of-

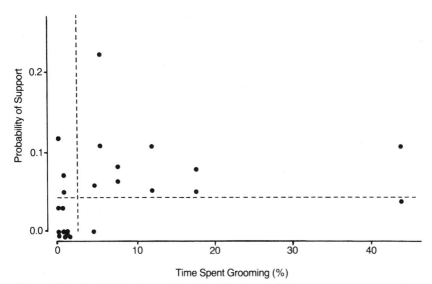

Figure 18   Probability that a female will give coalitionary support to another female of her unit, plotted against the percentage of time that the two females spent interacting with each other. Dashed lines indicate median values on each dimension. (Reproduced from Dunbar 1980b, Fig. 4).

fered by a more dominant (and hence aggressive) coalition partner. Figure 18 shows the frequencies with which females in three reproductive units came to the support of other females of their unit when those females were involved in agonistic encounters (as either aggressor or aggressee) with members of neighboring units; the abscissa gives the percentage of available social time that the two females spent interacting. It is evident that a female is likely to support any female with whom she interacts socially, and will not support a female with whom she does not groom (median test: $\chi_1^2 = 6.042$, $p<0.05$). Because all the females with whom she interacts are likely to be her close relatives (Dunbar 1983b), it is clear that a female will only support close female relatives. Indeed, she will do so even if they do not groom all that frequently with her.

Although there was a correlation between the relative frequencies with which pairs of females supported each other (Fig. 19: linear regression, $t_{10} [b = 0] = 34.962$, $p<0.001$), the more aggressive member of the pair supported the less aggressive significantly more often than *vice versa* ($t_{10} [b = 1] = -44.496$, $p<0.001$). Thus, a naturally low-ranking female gains in a strictly non-reciprocal way by forming a coalition with a more dominant female because she then ranks imme-

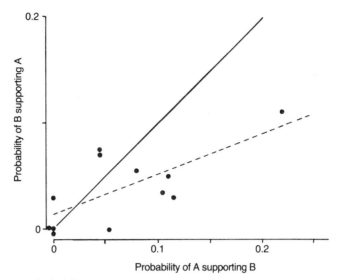

Figure 19 Probability that female A will support female B, plotted against the probability of B supporting A (where A is the more aggressive of the two). The broken line gives the least squares regression line; the main diagonal (solid line) indicates equal support rates. (Reproduced from Dunbar 1980b, Fig. 5).

diately below her ally (see Dunbar 1980b). Coalitions of this kind are particularly important to a female, for they raise her rank above that which she would otherwise hold at that stage of her life. A coalition does not, however, influence the rank of the more dominant member: she "gives away" aid with little prospect of gaining an immediate return (see Fig. 19). This is because a less aggressive female does not contribute substantially to a female's ability to dominate other members of her unit during her prime (because, with harem sizes of only 4–5 females, the numerical size of a coalition is unlikely to be a significant factor in any given encounter from a naturally high-ranking female's point of view). For a less aggressive (and hence naturally low-ranking) female, this aid has the effect of increasing her age-specific birth rate, thereby giving her a greater lifetime reproductive output than she would otherwise have gained. Simulation has shown that coalitions formed between mothers and daughters allow a female to maintain a higher rank for a longer period of time. Figure 20 compares the lifetime rank trajectory of the average female under conditions of no coalition formation (the constraint-free condition) with the trajectory she has if she

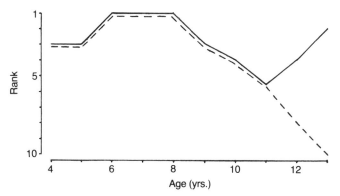

Figure 20 Estimated mean dominance ranks in each year of life for the average female if she forms lifetime coalitions (solid line) and if she does not (broken line). Based on results of the simulation given in Table A.1.

forms coalitions with her mother and daughter(s). (The general model used to determine ranks is given in Appendix A.1.)

We can use the regression equation derived from Table 6 to determine the lifetime reproductive outputs of females who form such coalitions and those who do not by iteratively applying this equation to the appropriate rank trajectories given in Figure 20 (for details, see Appendix A). In doing so, we need at the same time to correct the rank-specific birth rates to take account of the variation in fertility with age (see Fig. 13). The easiest way to do this is to linearize the relationship shown in Figure 13. Reapportioning the 8–12 year old females between the two adjacent age classes yields an equation for birth rate on age of:

$$y = 0.622 - 0.037x$$

(with $r^2 = 0.963$), where x is the female's age in years. Adjusting the rank-specific birth rate by the ratio of the female's age-specific birth rate relative to that of a female of median age, the expected lifetime reproductive outputs are 4.841 and 4.556 offspring, respectively, a net gain of 6.2% in favor of coalition formation. The greater part of this gain derives from the fact that coalitions dramatically improve the birth rates of older females, thereby offsetting the age-dependent decline in fertility. Note that the percentage gain is close to the 6% gain estimated by Dunbar (1980b, p. 263). As noted in Chapter 6, that estimate was obtained using a regression equation for converting rank into birth rate that was based on birth rates averaged over females of all ages, thereby in effect taking into account the variance in the age-specific birth rate. The two models do differ in one important respect, however. The present model considers the problems from the point of view of the *average* female (born when her mother is 8.3 years old) maturing into the average-sized harem (4 reproductive females); the model given in Dunbar (1980b) considered only the case of the first-born offspring (born when the mother was 4 years old) maturing into a unit of only 3 females. That the two models yield closely comparable results despite significant differences in initial conditions suggests that the general approach is quite robust.

EVOLUTION OF COALITIONARY BEHAVIOR

Because alliances are formed between relatives, we might infer that such behavior has evolved through kin selection (see, for example, Kurland 1977, Silk 1982) rather than through gains in personal fitness as suggested by Dunbar (1980b). In order to assess this possibility, we need to determine the relative gains in inclusive fitness (*sensu* Hamilton 1964) that derive from increments in a female's personal (i.e. direct) fitness and from increments in the fitnesses of collateral relatives. Note that the gain in a female's inclusive fitness is not, as is sometimes erroneously supposed, some proportion of the relative's *total* reproductive output, but only that proportion of the *increment* in this total output due to the female's altruistic behavior (Dunbar 1983a; see also Grafen 1982).

To estimate the female's gain in personal fitness, we need to determine the gain in her personal reproductive output that results from forming coalitions (i.e. her profit in excess of the output she would have had if she did not form coalitions), together with the gains obtained by her daughters through her influence on their ranks. Her gain through kin selection, on the other hand, derives from the extra offspring produced by her own mother and her sisters that result from her coalitionary support. Since she can influence the ranks (and hence birth rates) of female relatives only, the contributions made by sons and brothers need not be considered.

A female is in her physical prime between 6 and 10 years of age, and only during this period of her life can she influence the ranks of her relatives. Consequently, we need to consider only that portion of these various relatives' lives that falls within this period of her own life: thereafter, either she gains rank from them and/or they are responsible for their own rank determinations. Since the average female is 8.3 years younger than her mother, her mother's lifespan does not overlap with the critical period when she can give her mother rank support: only a first-born daughter can do this. The female's living sisters will, on average, be 3.5 years older and younger than she is, so that she will be able to influence their ranks when they are 9–13 and 4–6 years old, respectively. The female's oldest daughter will be born when she is 5.5 years old, and will mature into the adult cohort when she is 9.5 years old: the daughter will thus be able to benefit from the female's rank support only when she is 4 years old. The female's second daughter will be born after the period when the female can influence her rank. (Mean age differences between females of different kinship are determined in Appendix A.1.)

Since fitness, strictly speaking, is a measure of the number of genes contributed to some arbitrarily distant generation (see Dunbar 1982b and references therein), a more precise estimate of the contribution made by each of these individuals to the female's inclusive fitness will be given by the numbers of offspring born to the extra offspring that each individual gains from coalitionary support, assuming in this case that males and females produce equal numbers of offspring on average. (Note that, while the female can influence the ranks of female relatives only, the additional offspring, whether male or female, gained by these relatives contribute to the female's inclusive fitness.) For these purposes, I have assumed that the number of second-generation descendants produced in a lifetime by each extra offspring gained by the recipient individuals is that produced in the non-coalitionary condition (i.e. 4.556: see p. 59). The precise value is unimportant, since it really acts only as a scaling factor to give output values that are not too small for illustrative purposes: fitness, being a relative measure, is of course unaffected.

In making these calculations, we must bear in mind that an individual benefits through kin selection only by a proportion of the extra descendants gained by a relative, the proportion in each case being set by their coefficient of relationship, r (Hamilton 1964). Those extra offspring she gains through coalitionary support from her daughters during her old age will contribute 0.25 of their offspring to her fitness. On the other hand, her own support of her first-born daughter will yield a number of extra grandchildren, and the offspring of these will contribute 0.125 genes each to her fitness. Since daughters have a small likelihood of sharing the same father as their mother, the extra genes acquired through the paternal line need to be considered too. With an average tenure for harem males of around 5.25 years (see Chapter 13), the probability that a female will have the same father as her sister next in age will be:

Table 7

Mean difference in age between the average female and various female relatives, together with their mean coefficient of relationship weighted for the likelihood that they share the same father (see text for details).

Relative	Age Difference (years)	Mean $r_{\bar{x}}$
Mother	8.0	0.543
Living sister	3.5	0.414
Extra[a] sister	5.0	0.367
Extra niece	8.5	0.253
Eldest daughter	9.5	0.578
Extra daughter	13.0	0.516
Extra granddaughter	8.5	0.175

[a] Born as a consequence of coalitionary support.

$$P_f = (1 - 1/5.25)^x,$$

where x is the number of years separating the birth dates of the females concerned. The mean coefficient of relationship weighted by the probability of sharing the same father in any given case is:

$$r_{\bar{x}} = r_f \cdot P_f + r_m (1 - P_f),$$

where r_f is the coefficient of relationship if they share the same father and r_m is the coefficient of relationship if they are related only through the maternal line. Table 7 gives the mean age differences between the average female (the third-born offspring of her mother) and various key relatives, together with their weighted coefficients of relationship.

The number of extra offspring gained by each of the beneficiary females as a result of the subject female's coalitionary support is the difference between the number of offspring produced under the constraint-free condition and the number produced through coalition formation during the period of the subject female's influence. This can be determined by iteratively applying the regression equation for Table 6 (taking into account the correction for age-specific fertility derived from Fig. 13) using the algorithm outlined in Appendix A.2. In each case, the gain in rank was computed from the appropriate sections of the demographic trajectory of the average harem (see Table A.1 of Appendix A). In all cases except that for her mother, the period of the female's influence on a relative's rank comes from the midpoint of the trajectory. To standardize the gain base, therefore, the mother's gain was taken to be the average of the gains she would have acquired if

Table 8
Mean coefficient of relationship ($r_{\bar{x}}$) and gain in inclusive fitness (Δf) obtained from forming coalitions by females of different birth rank.

| Fitness Gain Through: | Birth Rank of Female | | | | | |
| | First | | Third[a] | | Fifth | |
	$r_{\bar{x}}$	Δf	$r_{\bar{x}}$	Δf	$r_{\bar{x}}$	Δf
Self	0.258	0.5019	0.258	0.3350	0.258	0.3542
Daughters	0.087	0.0582	0.087	0.0582	0.087	0.0873
Mother	0.194	0.2870	0.184	0.0086	0.150	0.0000
Sisters	0.154	0.0217	0.127	0.0741	0.154	0.0087
Personal fitness		0.5601		0.3932		0.4415
Collateral fitness		0.3087		0.0827		0.0087
Total gain		0.8688		0.4759		0.4502

[a] Equivalent to the average female.

the period concerned was at the beginning and the end of the harem's demographic trajectory. The number of second-generation descendants produced in each case was obtained by multiplying each of the values so obtained by the average number of offspring that each individual is expected to produce over the course of its lifetime even if it does not form coalitions itself (i.e. 4.556). These were then converted into genes contributed to the subject female's inclusive fitness by multiplying them by the appropriate coefficient of relationship. In each case, this coefficient has been assumed to be half that of the relevant mother on the grounds that the age gap between the subject female and a beneficiary's second-generation descendant is so great that the likelihood of their sharing the same father is negligible.

Note that, in computing the female's fitness gains, no deductions have been made for the costs of coalitionary behavior that are, strictly speaking, required by Hamilton's (1964) model. These have been omitted for two reasons: first, it is extremely difficult to measure the costs in fitness units and, second, the rarity of physical contact in fights between females suggests that they are probably negligible (Dunbar 1980b; see also Chapter 5).

The results of these calculations are given in Table 8 for females of three different birth ranks.

By forming coalitions with her close relatives, the average female can expect a total gain in her inclusive fitness equivalent to 0.4759 genes, of which only 17.4% derives from gains in the fitnesses of collateral relatives. Thus, the greatest contribution to her inclusive fitness

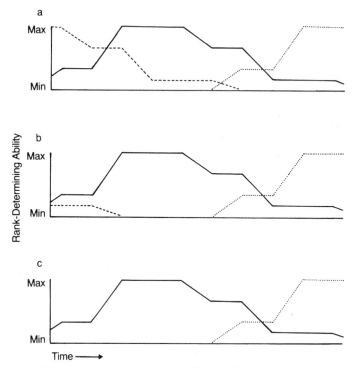

Figure 21 Schematic representation of the relative natural rank-determining abilities of a mother (dashed line), her daughter (solid line), and the daughter's oldest female offspring (dotted line) for the mother's (a) first-born, (b) third-born and (c) fifth-born daughter. Each female's status is shown between the time she matures (at the age of 4 years) and the mean age at death (14 years). The vertical scale is nominal.

does in fact stem from the gains in her own personal output that accrue over the course of her lifetime, particularly during the later years when she is able to gain reciprocated coalitionary support from her own daughters. However, because birth rank determines the extent to which she overlaps in age with her mother and her own oldest daughter (Fig. 21), the relative gains of the individual beneficiaries (and hence the female's gains in inclusive fitness) vary with the female's birth rank. The main point to notice is that a first-born daughter is optimally matched in age to maximize both her own and her mother's gains in personal fitness. A fifth-born daughter cannot expect to contribute to her mother's fitness at all. On average, a female gains through two consecutive periods of delayed reciprocal support: with her mother during the first half of her reproductive life and with her daughter(s) during the second half. The lower down the birth order, the less a female will gain from the first of these; conversely, the longer it takes her to produce a daughter, the less she will gain from the second. As a result, the proportion of the female's inclusive fitness that derives from increments in the fitness of collateral relatives declines from 35.5% to 1.9% as birth rank declines.

This is essentially a consequence of the particular life-history characteristics of these gelada and it is important to appreciate the fine balancing of reproductive lifespans and lifetime rank trajectories that is involved. As Figure 21 shows, a female's oldest daughter will, on average, enter the adult cohort (and hence become of interest for coalition formation) just after the death of the female's own mother. Thus, at no point is the female forced into a conflict of interest between the relative merits of her mother and her daughter as coalition partners. Were the natural rank trajectories or longevity different, she would encounter such conflicts and the optimal strategy would almost certainly be different as a result.

Although these calculations have been fairly crude, it is clear that the gains in personal fitness greatly outweigh the gains from kin selection. To offset the gains from kin selection, the average female would have to lose about three-quarters of her own gain in reproductive output: at the observed birth rate for very old females (0.1111 births per year: Fig. 13), this would mean dying about 2.38 years earlier than she would otherwise do (equivalent to a substantial 24% reduction in her breeding lifespan).

The model itself seems to be quite robust. This is largely a consequence of the fact that a female's gains in inclusive fitness derive mainly from improvements in her own birth rate and that of her mother. The relative contributions of these to her inclusive fitness are determined by the extent to which the offspring concerned are related to her and the extent to which their reproductive lives overlap. Inevitably, her own offspring are more closely related to her than are her sisters' offspring. Since the gain in direct fitness is substantially higher than that in collateral fitness, there is considerable leeway for error in the various estimates without serious risk of altering the qualitative results of the analyses.

The only real weakness in the model would seem to lie in the use of a specific lifespan, since this ignores all effects that arise in those cases where individual females live longer. In order to incorporate the full variance in life expectancy, however, we would have to evaluate the model for each possible lifespan and then adjust each outcome by its

Table 9

Increments in inclusive fitness that accrue to the average female and her eldest daughter from each female's influence on the other's ranks. The daughter gains rank support from her mother mainly during the first year of her reproductive life; the mother gains support from the daughter mainly during the last 2 years of her life (see Table A.1).

| | | Benefit to: | |
		Mother	Daughter
	Mother	0.0582	0.1725
From rank-gain by:			
	Daughter	0.3091	0.1042

likelihood of occurrence. Since a female's fertility declines with age, the contribution from the later years, once their low frequencies of occurrence is taken into account, will be vanishingly small. Consequently, it is doubtful whether such a complicated analysis would yield results that differ significantly from those obtained by considering the single case of the average lifespan.

Although the female's overall gain derives mainly from gains in her personal fitness, in each case the older individual in an alliance gives away a relatively small increment in her daughter's ranks during her prime while receiving a considerably greater benefit in return during her old age. Costing out the relative gains in inclusive fitness for the average female and her eldest daughter that derive from their respective influences on each other's ranks suggests that both benefit most from their own gains in personal fitness (Table 9). This is a consequence mainly of two factors: the weighting on reproductive rates against older females and the large difference in a female's coefficients of relationship with her own daughters and with the offspring of younger sisters (0.258 compared with 0.087). In other words, each maximizes her gain only if the relationship is reciprocal in the long run. On the other hand, since the costs of coalitionary support are probably small, the mother gains at least a minimum return through kin selection even if her daughter fails to survive long enough to fulfil her part of the compact.

In summary, a female's primary concern would seem to be to establish alliances with other females that would enable her to maintain as high a dominance rank as possible throughout her lifetime. Two questions remain to be answered. First, why isn't *any* female, related or unrelated, an acceptable ally? And, second, why don't females prefer to form coalitions with the harem male, given that he is invariably the most dominant member of the unit? Consideration of the second question will be deferred to the following section.

WHY FORM COALITIONS WITH RELATIVES?

One reason why unrelated females are less satisfactory as coalition partners is that coalitions are reciprocal only in the long term. Investment in a coalition is made, in effect, on the assumption that, over a lifetime, the female will gain at least an equivalent return. Moreover, the real benefit to a female comes during her old age. Such a reciprocal relationship would probably be viable only if it was based on an existing social bond of considerable strength. The mother-infant and sibling relationships are virtually the only ones that are suitable in this respect: they are of far longer duration and are psychologically much deeper than any other relationships that an animal might form during its lifetime. Thus, the *psychological* (i.e. proximate) basis for such a long-lasting relationship already exists in the familiarity between close relatives and the predictability of behavior that this engenders.

Another reason why an unrelated female makes a less than ideal ally is that the individual making the alliance gains only in personal fitness. An alliance with a close relative, on the other hand, gives her an additional benefit in terms of kin selection. Moreover, having a genetic investment in an ally probably makes it less likely that the coalition partner will renege on her return payments later on. These latter two considerations are probably of paramount importance in determining which particular individual(s) a female should ally with. Note that this does *not* mean that kin selection is the explanation for the evaluation of coalitionary behavior itself: there is a clear functional distinction here between doing something and whom you do it with. It pays to form coalitions with anyone, but if a choice is available, relatives provide an added bonus through kin selection.

This raises one further question: why are most alliances formed between mothers and daughters rather than between sisters (see Dunbar 1979b)? The main consideration militating against sisters is that it will be in the sister's interests to desert the alliance as soon as her eldest daughter becomes reproductively mature (at the age of 4 years). With the probability of each offspring being female equal to 0.5 and a 2.14-year birth interval, the sister's weighted mean age at the birth of her first daughter is 5.58 years (see Appendix A.1). The sister's age when this daughter enters the cohort of reproductive females will be $5.58 + 4.0 = 9.58$ years. Similarly, the weighted mean age gap between successive sisters is 3.41 years. Thus, desertion is likely to occur when the female is $9.58 - 3.41 = 6.17$ years of age if she allies with an older sister and $9.58 + 3.41 = 12.99$ years old if she allies with a younger sister.

With the critical age for maximal value as a coalition partner being 6–10 years, the older sister will begin to give coalitionary support when the female is 4 years old and will be of no assistance once the female's own rank-determining abilities reach their peak at an age of 6 years; more importantly, an older sister will be unable to help the female during the critical period of the female's life after the age of 10 years. The younger sister can begin to provide rank support only when the female is 9.5 years old, but, since the female still has approximately 1.5 years of her own critical period left at this point, the younger sister's full impact will not be felt until the female is 11 years old and will only last for the 2 years remaining until the sister's own daughter matures. In contrast, an alliance with her own oldest daughter is less prone to such conflicts of interest since the female's weighted mean age when her first granddaughter matures will be her own age at the birth of her first daughter (5.58 years) plus this daughter's mean age at the birth of her first daughter (also 5.58 years) plus the granddaughter's age at maturity (4.0 years), a total of 15.16 years. Since this exceeds the average age at death for females (13.8 years), a daughter is less likely to encounter conflicts of interest between her alliances with her mother and her own daughter. Consequently, a mother is less likely to suffer the risk of desertion than is a sister. Since her daughter will be able to offer maximal rank support when the female is $5.58 + 6.0 = 11.58$ years old, she will have approximately 2.5 years of rank support just at a point in her life-cycle when her own rank-determining abilities are going into rapid decline.

It might be argued, on the other hand, that a female should form coalitions with her daughter rather than her sister because the daughter's reproductive value (*sensu* Fisher 1930) will almost always be higher than that of her sister (see Schulman and Chapais 1980, Datta 1981). This argument fails in the present case, however, because the benefits conferred by the altruist are immediate rather than long-term and have their greatest effect on the reproductive outputs of older females. Furthermore, a sister is as likely to be younger than the female's daughter as she is to be older.

What makes this behavior advantageous for a female is the particular life-history characteristics of these gelada. It would be naive to assume that these results should necessarily apply to all other populations of gelada, let alone to all species of primates. Consequently, these results

Table 10
Correlations between the frequencies with which a harem male supported his females in agonistic interactions with neighboring units and the frequencies with which they groomed him.

Unit	Number of females	Kendall's τ	p^a
N2	4	0.000	0.620
N12	4	0.000	0.620
N21	3	− 0.333	0.667

[a] p-values are 1-tailed in a positive direction (see appendix to Dunbar 1978b).

need not be in conflict with Seyfarth's (1977, 1980) hypothesis relating dominance and grooming in a number of primate populations. Indeed, it is interesting to note that, while his model cannot easily account for the data from our wild populations, it does explain Kummer's (1975) data on captive gelada very satisfactorily (see Seyfarth 1977). This apparent discrepancy fits rather nicely with our current understanding of the rules governing the social strategies of female gelada. Kummer's groups differ from our wild ones in that none of his females were close relatives; in the absence of relatives with whom to ally, we would expect females who are unable to monopolize the male to form alliances with the highest-ranking females available to them (see Dunbar 1982c). The point is that, while general principles governing female behavior such as "maximize rank by forming coalitions" may well apply universally to all species, the expression of that principle in any given case will depend on the particular demographic characteristics of the population in question.

Male's Value as a Coalition Partner

There are two aspects to consider here. First, given that the male is invariably dominant to all his females, is he of any value as a coalition partner? Secondly, is it an advantage in terms of mating frequency and/or fertilization rate to be the male's main grooming partner?

The male does not appear to support his grooming partner more often than he does other females (Table 10; Fisher's procedure for combining probabilities: $\chi_6^2 = 2.722$, $p = 0.843$). There is equally no evidence to suggest that the male materially influences the female rank order (see Dunbar 1980b, Table 4), so his value as an ally is probably limited. Indeed, it is clear that a female's dominance rank determines her access to the male: as we saw in Chapter 5, the male's main social partner was usually the most dominant female in the unit who did not have a female grooming partner of her own. Thus, as far as partnering the male is concerned, a female is very much at the mercy of the dominant females in her unit.

There is, however, a more serious drawback to opting for the male as a coalition partner: a male's tenure as a harem-holder is relatively short. Estimates of tenure (see Chapter 13) suggest that, on average, a male can expect to hold a harem for between 3.5 and 7 years, depending on how he acquires it. This is well short of the female's expected breeding lifespan of $e_4 = 9.8$ years (see Fig. 11). Since an alliance needs to last a lifetime to show a real profit, the male cannot be a serious alternative to a female as an ally.

Nonetheless, for a female lacking relatives with whom to form an alliance, partnering the male may provide at least some short-term gains. By being able to count on the male's support for at least part of the time, a female may experience less harassment from dominant females than she would otherwise. This prediction can be tested by comparing the frequencies with which the male's main partner was threatened by the other females in her unit with the frequencies of threat made to a set of non-partner females matched for age, dominance rank and, where possible, harem size. Out of 7 comparisons, there were two ties (the male's partner being the dominant female in both cases); in the remaining units, the male's partner was harassed less often than the matched non-partner female (median rates of 0.046 and 0.513 threats per hour, respectively; Sign test, $p = 0.031$ 1-tailed).

By being the male's grooming partner, a female might also be able to reduce the amount of stress involved in having to bypass higher-ranking females in order to mate with the male. As the male's main social partner, she would spend up to 10% of the day in physical contact with him (sufficient time to allow several stress-free copulations). In baboons, ovulation is most likely to occur during the early hours of the day (Hendrickx and Kraemer 1969); consequently, since most social activity occurs in the first 2–3 hours of daylight, copulations during social time are optimally placed to ensure fertilization with the minimum delay. Copulations do in fact occur more frequently during the morning than during the afternoon (mean rates of 0.447 and 0.225 per hour, respectively: data derived from focal samples of individual oestrous females in 1971–72; see also Mori [1979a, Fig. 7.1]). In addition, it is generally true that the greater the frequency of copulation, the more likely it is that fertilization will occur on a given cycle (at least,

Table 11
Mean frequencies with which the male mounted and ejaculated for females who were and who were not his main grooming partner.

	Dominant Females[a]				Subordinate Females[a]				p[b]
	Ptr[c]	n	Nptr[c]	n	Ptr	n	Nptr	n	(1-tailed)
Mounts (n/hr)	0.445	4	0.473	3	0.420	4	0.438	3	0.155
Ejaculations (n/hr)	0.205	4	0.183	3	0.183	4	0.090	3	0.050
Ejaculations/ mount (%)	46.1	4	38.7	3	43.6	4	20.5	3	0.112

Source of data: 14 oestrous females from 7 reproductive units sampled in 1974–75.
[a] Dominant = ranks 1–2; subordinate = ranks 3–5.
[b] Wilcoxon's test comparing partner and non-partner females matched for age and dominance rank.
[c] Ptr = male's partner; Nptr = non-partner females.

up to the limit set by the male's rate of sperm production). Do partner females mate more frequently than non-partner females of the same rank? Or, more importantly in evolutionary terms, do partner females have higher reproductive rates than non-partner females of the same rank and age?

To answer these questions, data on oestrous behavior (summarized in Dunbar 1978b) were examined to determine the frequencies with which males mounted and ejaculated with partner and non-partner females of different dominance rank (see Table 11). Although only one of the differences is statistically significant, partner females were mounted and obtained ejaculation more often than non-partner females, and this is true for both high-ranking and low-ranking females separately. The likelihood of ejaculation per mount is also very much higher for partner females than for non-partner females (overall means of 44.9% and 29.6% of mounts, respectively). Thus, by being the male's partner, low-ranking females may have to work less hard to become fertilized than would otherwise be the case. The small difference in mating rate gained thereby is reflected in a comparable difference in reproductive output. The mean number of living offspring younger than 4.25 years of age for 11 partner females was 2.36, compared with 2.18 for a set of non-partner females individually matched for age, dominance rank, and harem size. Eight of the 11 partners had as many or more offspring than their matched non-partner female, and only 3 had fewer. Although none of these differences is statistically significant, they are all in the same direction, suggesting that a female may be able to gain a small advantage from allying with the male that at least partially offsets the disadvantages incurred by not having a related female ally. Applying Fisher's procedure for combining probabilities from independent tests, we obtain $\chi_4^2 = 10.352$ ($p = 0.035$) by combining ejaculation rate with number of offspring. (Ejaculation rate seems the best measure to use since it has a more direct bearing on fertilization; however, if we use the rate of mounting instead of the ejaculation rate, we obtain $\chi_4^2 = 8.089$, which only just fails to reach significance at the $p = 0.05$ level.)

In summary, the male is of less value as a coalition partner than a female relative. Nonetheless, for a female who lacks such a relative, he is better than nothing. Indeed, he is probably better than an unrelated female as a partner, since such a female is likely to renege on her alliance when her own daughter matures (perhaps as soon as a year or two later). The male, on the other hand, is unlikely to do so, and the female can keep up or terminate the relationship at her own convenience. This is the most plausible explanation of why females do seem to compete for the male. It would also explain why a male's partner continues to groom with him long after he has been displaced as haremholder by another male (see Dunbar 1983c): his value as a coalition partner is not altered by the takeover because he remains dominant to all the females of the unit.

Desertion: The Last Resort

A female locked into low rank can expect to remain low ranking for the rest of her life, the vagaries of the life-history variables aside. Since females normally remain in their groups for life, and high-ranking females breed faster than subordinates, a female's rank will decline with age as successive daughters of higher-ranking females rise into the female hierarchy. A female in such a situation has one final option: to desert her male and join another. Providing she joins a unit with fewer females than the one she came from, her dominance rank will necessarily be higher, even if she is still the lowest-ranking female in the new unit.

Although this sounds a sensible thing to do, females hardly ever leave their natal units. In more than 3 years of field work on several bands at Sankaber and Gich (equivalent to approximately 1500 animalyears of observation) only one female has ever been known to change units (see Dunbar 1980a, Ohsawa and Dunbar 1984). Yet the usual selection pressures militating against transfer between groups, such as increased predation risks (see Harcourt 1978, Frame and Frame 1977), are absent: females could transfer between units of the same band (or even those of different bands) without having to leave the safety of the herd. This is highlighted by the fact that hamadryas females, living under analogous social conditions but subject to very much more severe ecological pressures, regularly transfer from one reproductive unit to another (see Sigg et al. 1982).

Why will gelada females not desert their natal units, even when it might be in their interests to do so?

No direct evidence is available to answer this question, but comparisons of captive and free-ranging groups give some indication as to why this might be. Kummer (1975) found that, in contrast to the situation in wild populations, the females of his convened groups readily deserted their male when a second male was added to the compound. The main difference between the two sets of groups is that the females in Kummer's groups were all unrelated, whereas in the wild all the females of a given harem are related to each other at least to some extent (see Dunbar 1982c, Shotake 1980). Females, it seems, simply do not like to leave their relatives, even the more distant ones.

There are three probable reasons for this.

First, females separated from their units are objects of considerable interest to the bachelor males of the all-male groups. We have twice seen males from all-male groups interact with females separated from their units, on one occasion trying to groom with the obviously very nervous female and on the other actually mounting the female. The females were rather at the mercy of the males, who outnumbered them. Moreover, the sight of a female interacting with all-male groups is likely to precipitate frantic herding by her harem male, probably culminating in fighting between him and the bachelor male. On a number of occasions when individual females strayed more than 100 m from the rest of their units, the harem males were evidently very concerned (for examples, see Dunbar and Dunbar 1975, pp. 36,39).

The second factor militating against a female leaving her natal unit is that it is by no means easy to enter another reproductive unit, even a small one. Kummer (1975), in his experimental manipulations, noted that a female joining an existing group had to establish a relationship with the dominant female before she was allowed (by the dominant female) to interact with the male. In the one case of female transfer observed in the wild, the incoming female likewise established a relationship with the (probable) dominant female before she interacted with the male (see Dunbar and Dunbar 1975, pp. 100–102). Transfer to a new unit is thus likely to be a lengthy and difficult process, and there is every reason to presume that attempts at transferring will at least sometimes be unsuccessful. We observed harem males trying to capture "loose" juvenile females wandering near their units on two occasions, both times without success. On two other occasions, attempts by juvenile females to associate with (and presumably join) particular reproductive units proved unsuccessful in that they aroused no interest from the harem male (see Dunbar and Dunbar 1975, p. 100). At Gich, attempts by a harem male to integrate the females of two different harems ultimately failed because of resistance by the two groups of females (Mori 1979c).

Finally, and perhaps most importantly, the protection afforded by groups of related females against generalized harassment from other females in the very large herds that characterize the gelada is likely to be a significant factor militating against a female deserting her natal unit. A female who joins another unit will be joining a group of females who are, in effect, unrelated to her; their willingness to support her during altercations both with neighboring units and among themselves may be greatly reduced in consequence.

Although a female cannot (or will not) desert her male by moving to another unit, she can desert him in two other rather different ways without (at least initially) leaving her natal unit. These are (1) by attaching herself to a follower male if and when one joins the unit and (2) by attaching herself to a male who attempts to take over the unit (see Dunbar and Dunbar 1975).

There are good reasons to suspect that females would consider these to be separate options. Even though in both cases the female can expect to gain by becoming the new male's partner, by joining a follower she has the additional advantage that she will in due course belong to a smaller unit, because, sooner or later, the follower will separate from the parent unit taking with him those females that he has "kidnapped" from the harem-holder.

On average, 4 females were taken by the follower when units underwent fission, while the harem-holder retained an average of 4 females also ($n = 3$ units). Thus, females who transfer their allegiance to the follower can expect a substantial increase in rank (and hence reproductive rate) when the unit undergoes fission. In the extreme case, the bottom-ranking female would rise from an average rank of 8 in the original unit to one of 4 in the follower's new unit. This would mean an increase in her expected birth rate from 0.229 to 0.419 births per year.

On these grounds, we might expect females to prefer a follower to a takeover male. The opportunity for choice, however, will be rare, for neither event occurs with any great frequency in a female's lifetime. Nonetheless, females who have already opted to transfer allegiance to a follower may later have a chance to desert again in favor of a male attempting to take the unit over. Do such females show any evidence of preference one way or the other? Two units that had followers were taken over by a total of 3 males (H39 in 1972, N5 by 2 males in succession in 1975). In both units, the follower's female (in the case of N5, the female mated exclusively with the follower) vigorously resisted attempts by the new males to take her over. The case of H39 a few days after its takeover is illustrated in Figure 38 of Dunbar and Dunbar (1975): this relationship survived for more than 2 years after the takeover (see Dunbar 1979b, Fig. 1). In the case of N5, both males who took over the unit made repeated efforts to detach the female from the follower. The follower's response on each occasion was to cower, screaming at any sign of threat from the two older and rather larger males, whereas the female's response was to attack the new males and drive them off even though she was much less than half their size. I interpret this behavior to mean that the female had made her decision and intended to stick to it no matter what the new males tried to do. In both cases, the new males gave up trying to herd the female after a few days of fruitless effort. This is rather dramatic, if anecdotal, evidence in support of the hypothesis that females prefer a follower to a new harem-holder.

Which females are most likely to desert their male in favor of a new one? Perhaps the most obvious hypothesis is that, given the relationship between dominance rank and reproductive rate, low-ranking females should be the first to desert. Of four units that were taken over for which detailed observations are available, the dominance rank of the first female to desert is known in three cases. In all three, it was the lowest-ranking female who was first to interact with the intruder. The likelihood that the first deserters would be at least as low ranking as the observed females by chance is small (Fisher's procedure for combining probabilities: $\chi_6^2 = 11.983$, $p = 0.062$).

These results in general concur with those obtained by Kummer (1975) for his captive animals: he found that subordinate females were most likely to desert, though he explained this by the fact that, in his groups, the frequency of grooming with the male was a simple function of dominance rank. None of the females in Kummer's groups were related, which suggests that, in the absence of kinship bonds, dominance is the main factor governing a female's loyalty (see Dunbar 1982c).

Although low rank was the main characteristic of the first female to desert, this female did not necessarily become the new male's main grooming partner. The first female to desert and establish a relationship with the new male was often subsequently displaced by a more dominant female, and the new male's final grooming partner was typically a young adult, dominant, partnerless female (see Dunbar 1983c). This suggests that, if a female does want to improve her relative rank, attaching herself to a follower may be a more profitable strategy than trying to become the new male's partner following a takeover.

8

A Decision Model
of Female Reproductive
Strategies

The analyses in the previous chapter suggest that a female has six main
strategies open to her: (1) to form coalitions with her mother and
daughter(s), (2) to form a coalition with a sister, (3) to form a coalition
with a less closely related female, (4) to become the harem male's
social partner, (5) to become the new male's partner when the unit is
taken over and (6) to join the follower when her unit undergoes fission.
Can we evaluate the relative profitabilities of these various strategies
in order to say something about the optimal strategy choice for a female faced with the problem of low rank and its attendant low reproductive rate?

In this chapter, I develop a decision model for females in order to
try and answer this question. The analysis, however, is complicated by
the fact that the problem is essentially a demographic one: a female's
lifetime rank trajectory depends on who happens to be in her unit in
each successive year of her life. This means that we cannot derive a
simple algebraic model of the kind commonly found in sociobiological
applications. Instead, we need to work with a statistical simulation model
that tries to mirror the changing experience of the average female over
the course of her life. Although this inevitably means that the resulting
analyses seem to be somewhat crude and inelegant, they have the merit
of being realistic.

Strategy-Specific Profitability

The value of each of these strategies is likely to depend on the female's
age and reproductive prospects. Consequently, in order to assess their
relative merits, we need to consider females at different stages of
their life-histories. I shall do so for females who are at the start of their
reproductive careers (aged 4 years), those in their prime when their
own rank-determining abilities are at their peak (aged 7 years) and

those about to enter the period of decline (aged 10 years). Moreover, since the previous analyses have suggested that birth rank has a significant influence on a female's inclusive fitness, the analyses will be carried out for first-born, third-born, and fifth-born females at each age for the first three strategies. (Birth rank will not affect the gains derived from strategies that involve males as coalitionary or social partners.)

To gauge the relative value of each strategy, it is convenient to have a baseline against which the relative profitabilities can be compared. The obvious candidate is the constraint-free strategy (i.e. the condition under which no coalitions are formed and a female's rank at any given age is determined solely by her intrinsic fighting abilities relative to those of the other females in her unit). Using this as a baseline, I have estimated the percentage increase in inclusive fitness gained by pursuing each of the strategies for females of the three birth ranks at the three different ages. The results are given in Table 12.

The profitabilities of the first three strategies are determined directly by iteratively applying the regression equation for birth rate on dominance rank derived from Table 6 (adjusted for age-specific fecundity using the relationship derived from Fig. 13) to the appropriate age/birth-rank–specific lifetime rank trajectory, following the algorithm given in Appendix A.2. The lifetime rank trajectory of the average (i.e. third-born) female is given in Figure 22 for the four types of coalitionary strategy and for the contraint-free condition (no coalition formation). We know from the earlier analyses that the gain to a 4-year-old third-born female is 6.2% if she forms coalitions with her mother and daughter(s) (see p. 59); the gains for first-born and fifth-born females are 6.3% and 3.3%, respectively. To determine the corresponding values for older females, I have assumed that the female occupies the ranks appropriate to a non-coalitionary female until the year in which she makes her decision, after which she occupies the ranks appropriate to a coalitionary female. In all these cases, the female also gains in personal fitness through her daughters and in inclusive fitness through kin selection due to her influence on the ranks of her mother and sister(s). The relative gains in fitness obtained from each beneficiary are given in Table 8 for females of the three birth ranks who make their decisions at 4 years of age. The ratio of these values to her own gains (top line of Table 8) was used to adjust the female's personal fitness in order to estimate her total gain in inclusive fitness. Comparable values were calculated for females who make their strategic decisions at 7 and 10 years of age.

We know (from p. 68) that a female who forms coalitions with a sister sustains a loss in her reproductive output relative to what she

Table 12
Relative value of various strategies to females of different ages, measured in terms of the percentage increase in inclusive fitness gained over the constraint-free strategy of no coalition formation. For each of the first three strategies, the values are given for three different birth ranks.

Strategy	Birth Rank	Increase in Fitness (%) Age of female (years)		
		4	7	10
1. Coalition with mother/daughter	1	15.9	12.9	8.2
	3	6.5	6.2	5.7
	5	7.4	7.4	7.4
2. Coalition with sister	1	3.4	3.4	3.4
	3	5.6	0.1	0.1
	5	3.1	0.0	0.0
3a. Coalition with aunt/niece	1	0.0	0.0	3.5
	3	0.0	0.0	4.4
	5	0.0	0.0	1.7
3b. Coalition with unrelated dominant female	1	0.8	3.1	3.0
	3	0.0	0.0	0.0
	5	3.2	2.6	1.7
4. Become harem male's social partner		3.0	3.0	3.0
5a. Become new male's social partner after takeover		4.6	4.6	3.3
5.b Become follower's social partner		5.8	5.8	3.3
6. Join follower's harem at fission		16.1	11.5	10.3

would have gained had she allied with her mother and daughter(s), as a result of desertion by her sister when the sister's oldest daughter matures. The female's net gain in personal fitness was determined by assuming that she occupied a rank equivalent to that of a non-coalitionary female until the year in which she made her decision, then occupied the rank immediately below that of her sister until the time at which desertion occurred, thereafter reverting to the rank trajectory of a non-coalitionary female. If no sister was available to act as an alliance partner (either because the timing was such that, on average,

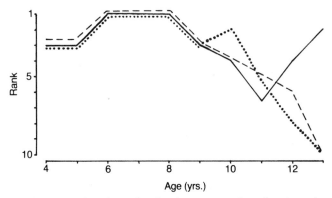

Figure 22 Lifetime rank trajectories for the average female when she forms coalitions with her mother and her own daughter(s) (solid line), her sisters (dashed line) and aunts and nieces (dotted line). Based on the analyses given in Appendix A.

no sister would be of the right age or because the sister's own daughter was mature, thus pre-empting the sister's coalitionary activity), the female was assumed not to be able to benefit from coalition-formation. Conversely, if more than one sister was available, the gains were computed separately for each possible demographic configuration and an average gain then determined. Because the female does not ally with her mother, her gain through kin selection is limited to that which accrues through her sisters: for the three birth ranks considered in the analyses, this turns out to be 6.3%, 21.5% and 2.5% of her gain in personal fitness for first-, third- and fifth-born females, respectively.

Two possibilities were considered for the third strategy (allying with an "unrelated" female), namely, allying with an aunt or a niece and allying with the most dominant female available in the harem at the time. The main disadvantage in both cases is that the alliance partner is likely to desert as soon as her oldest daughter matures. For those females who could form a useful alliance (i.e. one that increased their rank within the harem), the mean time to desertion was determined from the demographic configurations and, averaged across all ages and birth ranks, turned out to be 1.9 years. The female's gain from this strategy was calculated in the same way as for the second strategy (allying with a sister), except that, because a female does not contribute to her ally's rank after she has been deserted, there could be no gain due to kin selection.

It should be noted that, once again, no deduction has been made for the costs of coalitionary behavior when calculating the net gains in inclusive fitness. It has been assumed, as previously, that these costs are likely to be negligible (see p. 44).

The profitabilities of the three strategies involving male partners were calculated rather differently. We know (from p. 71) that a female gains an increase in reproductive rate of 8.3% as a result of reduced harassment and stress when she becomes the harem male's main social partner. However, she has this gain only during the period of the male's tenure as harem-holder and, in cases of takeover, during the period he remains in the unit as an old follower. Assuming that, on average, the male is halfway through his period of tenure when the female makes her decision, she will benefit for a period of only 3.6 years. (A male has tenure for between 3.5 and 7 years, depending on how he acquires the unit, and will remain active in the unit for about 2 years after being deposed in a takeover in the first case: see Chapters 13 and 15.) Since females of all ages will outlive the male's period of residence in the unit, they will gain by 8.3% over baseline only during the 3.6 years following the decision, while gaining no advantage during the remainder of their 10-year reproductive lifespan.

A female who becomes the new male's partner following a takeover can expect to gain the same advantage for the full 3.5 years that the male will, on average, hold the harem, plus the 2 years that he will remain in the unit as an old follower after being deposed. Thus, sub-adult and prime females can expect the male to be available for 5.5 years of their 10 reproductive years, while old females will gain during only the 4 final years of their lives. If the female becomes the young follower's partner, she will also presumably gain by 8.3%, but in this case the gain will be over the male's full reproductive lifespan of approximately 7 years.

Finally, females who join the follower when their unit undergoes fission gain from the fact that the harem size is thereby, on average, halved. To determine the net gain to such a female, I have assumed that the effect of joining the follower is to give the female a rank equivalent to half that of a non-coalitionary female from the year in which the decision was made, her ranks prior to that time being those of the non-coalitionary baseline.

The percentage gains over baseline for each strategy (given in Table 12) confirm that females should show a clear preference for forming coalitions with close female relatives, with mother/daughters being given priority over sisters. Birth rank has a marked effect in all cases, though in no case is the qualitative preference order altered. Females who cannot form such coalitions will generally prefer to become the male's partner rather than form an alliance with an unrelated female, although there is an age-dependent cline in the relative values of these alternative strategies: the younger a female is, the more profitable a partnership with a male becomes, while an old female may find an alliance with an unrelated female to her advantage. Younger females will usually find an alliance with a follower more advantageous than one with either a male who takes the unit over or with the incumbent haremholder.

Note that it will always pay a female, no matter what her age, to join the follower when her unit undergoes fission. The resulting reduction in harem size has a considerable impact on the reproductive rates of all females, even though the gain-rate declines with age. Note also that the harem splits exactly in half (within the limits imposed by the fact that relatives always stay together): this is the solution that will maximize the benefit to each female individually, and we can expect frequency-dependence to stabilize the observed values at around this level.

Changing allegiance to a new harem male during a takeover does incur serious risks for a female, since the other females of the unit actively attempt to prevent her from deserting, and fight, often viciously, among themselves once the issue between the two males has been decided (Dunbar and Dunbar 1975, p. 111). The fighting can be severe: on one occasion an infant died as a result. In addition, five females (three at Gich and two at Sankaber) underwent abortions following the takeover of their units (see Mori and Dunbar 1984). Taken together, the loss on the year's total reproductive output of 18 infants conceived into the sample of 7 units was a substantial 33% (with 95% confidence limits of 19–62%). To an old female with little chance of making up any losses that might be sustained, the marginal gains almost certainly do not outweigh the likely costs. Thus, old females should tend to view a change of allegiance with even less enthusiasm than is suggested by Table 12.

Note that a female is most likely to desert before her paracallosal sex skin has completed its color change from purple to pink. Dunbar (1977b) suggested that the delay in changing color might be an attempt to prolong the appearance of adolescence, thereby reducing the likelihood that the harem male would try to prevent a subadult female from deserting. If this is true, then it constitutes an attempt by females to keep their options open as long as possible. Although a juvenile female is of less interest than a mature female to a harem male (because of the risks of inbreeding), she is of considerable interest to a prospective follower (in particular) since, with a long reproductive life ahead of her, her value to the male is much greater than that of an old female who might die at any time.

Although these calculations are relatively crude, the differences are in general of such a magnitude as to make it unlikely that any tinkering with the values used in the computations would make a significant difference to the results. As we have already noted, the basic model of rank trajectories itself seems to be quite robust.

While a female may rank the strategies in the way suggested by Table 12, her options are in practice constrained both by the demographic context in which she happens to find herself and by her current rank in the harem. The dominant female in the unit might well prefer to ally with a female relative, but she will not necessarily have one available at the time. No fewer than three of the dominant (i.e. rank-1) females in the 11 sampled units did not have a mature female relative with whom to groom. The female's preference in this case will generally be to become the harem male's partner, at least until her eldest daughter matures. A lower-ranking female, however, may not have this option open to her: her choice will consequently lie between forming a coalition with an unrelated female and waiting for the unit to acquire a second male (either by takeover or by the entry of a young follower). It seems that a young female has little to gain by forming alliances with an unrelated female and will probably do better to concentrate her attention on her immature offspring pending the appearance of a new male.

Note that, since older females may find a short-term alliance with an aunt/niece to their advantage, some grooming dyads involving old and young adult females might have been erroneously assumed to be mother-daughter or sibling pairs. There were 9 such dyads in a sample of 38 grooming dyads (24%) in 14 reproductive units (see Dunbar 1979b, Table 8). However, alliance formation is a two-way process, and since young adult females gain no advantage at all from forming alliances with unrelated females (Table 12), it seems unlikely that many such dyads would arise in practice, however desirable an old female might find such an arrangement.

Waiting for a new male to join the unit will always be, at best, an opportunistic strategy. The rate of entry into units by males was approximately 0.21 per unit per year (see Table 46): a female can thus expect to see only two such entries during her 10-year reproductive lifespan. The frequencies of multimale units in the population suggest much the same rate of entry: about 20–25% of units had more than one adult male (see Table 2). Very low-ranking females will therefore tend to become trapped in a low-rank trajectory with little hope of escape.

Testing the Model's Predictions

A female is able to make strategy evaluations throughout her reproductive lifespan. Not only does this mean that she can continuously adjust her strategic decisions as the demographic context changes, but it also means that it will be difficult to define an optimal strategy that has universal validity. The optimal strategy will always depend on the particular demographic configuration of the female's unit, as well as on her age, birth rank, and dominance rank at the time. Note, in addition, that strategy choice is categorical (i.e. hierarchical), not stochastic. Consequently, the frequency distribution of the strategies in a population depends not on their relative profitabilities, but on the likelihoods of specific socio-demographic states.

The fact that we cannot predict an overall distribution for the various strategies which could then be compared with the observed frequencies makes it difficult to test the validity of these analyses. However, we can test particular predictions generated by the results given in Table 12, and these will serve as our evaluation of the reliability of the basic model and the analyses based on it. In most cases, the existence of a strong grooming relationship is taken as evidence of a coalitionary alliance (see Dunbar 1980b, 1983b).

There are six predictions we can make: (1) since females prefer alliances with close relatives, the number of females who do not form a strong grooming partnership with another female will reflect the likelihood that a female will not have any such relatives; (2) females will show a clear preference for mothers and daughters over sisters; (3) when the choice is available, females will groom most with the older of two daughters; (4) females who groom with the harem male will tend not to have a female relative in the unit; (5) females will only groom with a young follower if they have no female relatives with whom to groom; and (6) young females will be more likely to desert in favor of a new male than old females.

In order to test the first prediction, we need to know how many females are likely to have no female relatives with whom to interact. Using the analysis of the theoretical distribution of matriline sizes given by Dunbar (1983b, Fig. 4 and Table 3), we can determine the proportion of females who are likely to be left without a 10%-grooming-partner if (a) females groom only members of their own matriline and (b) each female can form only one such dyad. (This is similar to the method used by Dunbar [1983c, Table 6] to determine the likely distribution of the ages of females who could become the harem male's social partner.) In principle, this amounts to one female contributed by each matriline whose size is an odd number. On this basis, 14.7% of females would not have a relative free to groom with them. In the sample of 25 reproductive units, 36 out of 118 females (30.5%) were not members of a female grooming dyad. This differs significantly from that predicted by the theoretical distribution of matriline sizes ($\chi_1^2 = 23.322$, $p \ll 0.001$). However, the theoretical matrilines contain at least some aunt-niece dyads (in fact 12% of all dyads). Since aunts and nieces are of negligible value as allies (Table 12), females should treat these individuals as "unrelated." Excluding aunt/niece dyads increases the expected proportion of partnerless females to 26.7%: the observed distribution does not differ significantly from this ($\chi_1^2 = 0.877$, $p > 0.30$).

Note that this analysis has, in effect, confirmed two different predictions of the model. It confirms that females who cannot form strong grooming partnerships (and hence coalitionary alliances) with close female relatives will make no attempt to groom with other females. In addition, it also confirms the model's prediction that females will treat aunts and nieces as though they were unrelated.

The second prediction can be tested by comparing the relative frequencies with which females formed grooming dyads with mothers/daughters and with sisters. Of 38 grooming dyads in 20 reproductive units, 32 were between females whose ages differed by two or more age classes (hence likely to be mothers and daughters) (see Dunbar 1979b, Table 8). Only 6 involved females of the same or immediately adjacent age classes (who could thus have been sisters). If the dyads in this sample had formed at random within their particular units, 69.4% would have consisted of females close enough in age to have been sisters. The observed and expected distributions differ significantly ($\chi_1^2 = 3.891$, $p < 0.05$). Thus, females show a clear preference for mothers and daughters over sisters. In a captive group of known lineage, all grooming dyads were formed between mothers and daughters and none between sisters or aunts and nieces (Dunbar 1982c).

The third prediction was that a female will show a clear preference for her older daughter over her younger one. In the 25 reproductive units that were sampled, there were 9 females who had two living postpuberty daughters and one female who had three. Of these 10 females, 8 groomed the older daughter more than the younger (binomial test, $p = 0.055$ 1-tailed; mean time spent grooming was 24.2% with the older daughter and 15.9% with the younger). Of the two exceptions (which included the female with three living daughters), two of the older daughters had strong (i.e. >10%) grooming relationships with the harem male, while the third groomed extensively with her own first-born daughter (aged 2.5 years and therefore on the verge of entering the cohort of reproductive females).

The fourth prediction was that females will groom with the harem male only when they do not have a close female relative available. In the sample of 25 units, the male's partner had no female grooming partner in 14 cases (56%). This is significantly more than would be expected if the male's main partner were selected at random from among the females in his unit (30.5% of females did not have female grooming partners: $\chi_1^2 = 7.957$, $p<0.01$); it is also significantly more than would be predicted by the theoretical distribution of matriline sizes (when 26.7% of females would not have a close relative to groom with: $\chi_1^2 = 10.866$, $p<0.001$). Moreover, in only 5 of the 11 cases in which the male's partner also had a female grooming partner was there another female in the unit who did not have a female grooming partner (and two of those had strong grooming relationships with their unit's follower male). If only those units in which at least one female lacked a female grooming partner are considered, the male's partner was such a female in 14 out of 19 cases (74%). This is, again, very much higher than would be predicted if the male's partner were chosen at random from among the females of the unit.

Since females maintain their relationships with their harem males even after the males have been deposed as harem-holders (Dunbar 1983c), old followers could be counted as harem males for present purposes. Of 8 old followers in the sampled units, a female groomed extensively with the old male in 5 cases. In 4 of these, the female had no other grooming partner. In the 3 cases where no one groomed the old male, all the females had female partners (or groomed with the harem male). This distribution is significantly different from a random distribution (Fisher exact test, $p = 0.025$ 1-tailed).

The fifth prediction was that females would groom with the follower only if they had no female relative with whom to interact. Of 5 units with young followers, the follower had a strong (i.e. > 10%) grooming relationship with a female who did not groom with any other females in the unit in 4 cases; in the fifth unit, the follower had no female interactees, and all the females in that unit had at least one female 10%-grooming-partner. With such a small sample size, this is as close as one can get to a significant result (multinomial test, $p = 0.10$ 1-tailed).

The final prediction was that young females will be more willing to desert than old ones. Of 4 takeovers observed in sufficient detail, the first female to desert was a subadult in three cases, the fourth being a young adult. We can test this hypothesis in a different way by considering female grooming dyads (i.e. the bonded grooming partnerships) and asking whether there is any consistent tendency for the younger of the two females to form stronger bonds with the new male. Of 11 dyads in the same 4 units, the younger female established a stronger bond with the new male than the older in 9 cases (Sign test, $p = 0.033$ 1-tailed).

In summary, all six predictions have been upheld by the data, a result that is unlikely to occur by chance (binomial test, $p = 0.016$). This suggests that the analyses of female socio-reproductive strategies are generally valid, despite the conceptually simple nature of the model used to study them.

Is Fitness Equilibrated?

Whether or not the outputs of the various reproductive strategies equilibrate in the long run is virtually impossible to say. Taken at face value, they cannot do so, for they form a clear hierarchical sequence of preferences with rather little opportunity for other considerations to allow the losses on the poorer alternatives to be offset. Thus, strategies that are *prima facie* inferior do not seem to function as *bona fide* evolutionary options, but rather as "best of a bad job" alternatives to the optimum strategy (see Dunbar 1982a). As such, they serve to buffer the female against the serious consequences she would otherwise incur whenever she does not have close female relatives in the unit, but they do not allow her to make up that loss in full.

Nonetheless, it may be that the vagaries of the demographic processes are such that, on average, most females experience decision crises at some point in their lives. By optimizing their age-specific strategy decisions, females may able to equilibrate their fitnesses. If fitnesses could be shown to be equilibrated, it would be evidence to support the suggestion that females are able to optimize their strategic decisions over the course of their lifetimes in order to minimize the loss in fitness due to untoward demographic events.

In practice, it is not possible to put this to the test directly. To do so would require knowledge of the lifetime reproductive outputs of a large sample of females (in order to control for the effects of differential longevity) as well as knowledge of their genetic relatedness (in order to be able to determine the benefits derived through kin selection). Nonetheless, a partial test can be made by considering data on the

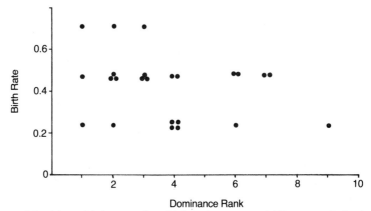

Figure 23   Mean birth rates for old females over a 4.25-year period, plotted against their dominance ranks. The data were obtained by using the number of offspring less than 4.25 years old that each female had to estimate the mean birth rate over that period. The estimates are not corrected for post-natal mortality. Source of data: 12 reproductive units sampled in 1975.

number of surviving offspring that females of different ranks were known to have. We can use these data to determine whether the rank-dependent birth rate averaged over an extended period of time is closer to a slope of $b = 0$ than is the instantaneous birth rate (as given in Table 6). This can only be a partial test, however: while a positive result supports the hypothesis, a negative result is equivocal for at least three good reasons. First, the data derive only from part of the female's reproductive lifespan (in fact, 4.25 years out of an average expectation of 10 years). Second, the effects of differential post-natal mortality will not be wholly taken into account (though the demographic data suggest that this effect is likely to be small: see Chapter 4). Finally, no account will have been taken of the possibility that a female could offset losses on personal reproduction by contributing to the reproductive outputs of her close relatives.

With these provisos in mind, each old adult female's mean birth rate was determined from the number of offspring less than 4.25 years old that she had. (Subadult and young adult females have not been considered here because they were too young to have been sampled over the full period.) The results are plotted against dominance rank in Figure 23; a least-squares linear regression is not significantly different from zero (t_{22} [$b = 0$] $= -1.403$, $p = 0.349$ 2-tailed). More importantly, the slope is significantly less steep than that predicted by the relationship between instantaneous birth rate and dominance rank derived from

Table 6 ($t_{22}[-0.0208 = b = -0.0476] = 1.825, p < 0.05$ 1-tailed). In other words, the relatively large differential in instantaneous birth rates is evened out in the longer term.

The marked reduction in the rank-specific birth rate when averaged over a period equivalent to nearly half the female's expected reproductive lifespan suggests that the differential may well be reduced to zero over the whole lifespan. The rate of change in the slope of the regression as the sampling interval is extended from 0 to 4.25 years suggests that a slope of $b = 0$ would be achieved after approximately 7.5 years (about three-quarters of the average reproductive lifespan). (Adjusting the birth rate to compensate for neonatal mortality would yield a zero slope after 8.3 years.) This suggests that females may in fact be able to equilibrate lifetime reproductive output, and that they can do so without reference to the fitness of relatives. If so, it implies that increasing the fitness of collateral relatives in order to benefit through kin selection is not a strategy that is regularly used to counterbalance the losses in personal fitness incurred through low rank.

The results in general tend to suggest that females attempt to maximize their instantaneous reproductive rates on a continuous basis in order to maximize lifetime reproductive output, with no attempt being made to trade short-term losses for long-term gains. This may be either because such give-and-take on a strategic scale is not possible for demographic reasons or because the females are incapable of making decisions of such sophistication and complexity.

9 The Female's Tactical Options

Within the general constraints imposed by long-term strategic considerations, females do have some freedom of movement in terms of the tactics they can pursue to offset losses in reproductive rate. There are three groups of tactical options open to females: behavioral tactics during oestrus (aimed at improving the probability of impregnation on each oestrous cycle); reproductive tactics (aimed at increasing the female's contribution to the species' gene pool); and tactics involving differential parental investment (aimed at maximizing an offspring's chances of survival).

Such tactics are unlikely to be equally attractive to females of all ages, however. Thus, attempts to step up the reproductive rate may be less profitable for older females because of the negative effects on mother and infant survival rates that are likely to ensue. Although an old female might be able to increase her reproductive rate, by doing so she might forfeit her chances of rearing the one last offspring she could otherwise expect to rear. Such a tactic would thus be counterproductive. Younger females, on the other hand, have more to gain and less to lose since they are physically stronger and have time to make up any losses they might incur. As a result, we would expect young females to pursue whatever tactical options allow them to offset strategic losses more frequently than older females.

Behavioral Tactics of Oestrus

The female's problem during oestrus is to get fertilized. To do so, she must, first, ovulate and, second, persuade the male to mate with her at around the time of ovulation. Although high-ranking females seldom interfere directly with the copulations of other females (but see Mori 1979b), low-ranking females must nonetheless cope with both increased harassment from them during oestrus and with the physical presence of these females between them and the male during non-social periods. One way in which a low-ranking female can minimize these psychological effects is to attract the male's attention, thus making him come to

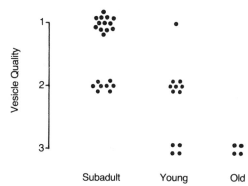

Figure 24   Distribution of scores for oestrous beading quality for females of different age classes (1 = neat, uniform necklace of vesicles; 3 = variable sized vesicles, scattered randomly across chest patch). Based on data given in Dunbar (1977b, Table 4).

her rather than vice versa. This can be done either by advertising her sexual state by some visual signal (as *Papio* baboons do with their sexual swellings, for example) or by doing so vocally. Both types of signal are found in gelada, the first in the vesicles around the female's sex skins, the second in the solicitation call given by females during oestrus (see Appendix A of Dunbar and Dunbar 1975).

There are good demographic reasons for a subordinate female, in particular, stepping up her reproductive rate early on in life. First, she cannot guarantee living to an old age, and, even if she does, the costs of reproduction are likely to be much higher for an old female. Both the risk of defects in offspring and the risks of perinatal mortality increase with maternal age (for data on humans, see Logan 1959). Second, in expanding populations, offspring born earlier in the mother's life contribute more to the species' gene pool than do offspring born later on (see Hirschfield and Tinkle 1975, Stearns 1976).

Visual Signals

Gelada females show a marked decline in the quality of their oestrous vesicles with age, those of old females tending to be irregular in size and scattered across the chest patch (in particular) rather than forming a neat ring around the edge (Fig. 24). In addition, the sex skins of juvenile females develop a marked puffiness during oestrus (Fig. 25). Irrespective of the *physiological* explanation for these two effects, it is clear that they do act as signals of the female's receptivity. Indeed,

Figure 25   Chest patch of a juvenile female illustrating the swelling common among pubescent females during oestrus.

they could be considered as super-stimuli in the classic ethological sense. At present, we do not know whether females of different dominance status can effectively influence the size and quality of these signals; on balance, it seems unlikely.

VOCAL SIGNALS

Females in oestrus repeatedly give a call that is quite distinctive and specific to this reproductive state. This call, termed the "solicitation call," is given throughout the oestrous cycle, but is significantly more frequent during and just prior to ovulation than during the menstrual part of the cycle (Fig. 26). Seven females were sampled to determine

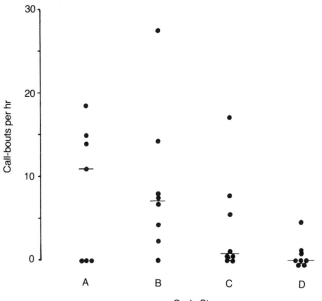

Figure 26   Rates with which females gave solicitation calls during each stage of the oestrous cycle. Bars indicate median values for each period. Following Dunbar (1978b), stages A to C refer, respectively, to early, full, and late oestrus; stage D refers to the period around menstruation. Source of data: 8 females sampled during 1974–75.

the contexts in which they called (Table 13). Of the total sample of 833 call-bouts, nearly 40% were given while the female was feeding, with no obvious stimulus that could be said to have elicited the call; 35% were given while the male or the female were moving, 12% while they were interacting with each other, 8% when the female looked at the male during a feeding bout and 4% while the male was involved in an agonistic encounter with either an all-male group or another harem male.

Solicitation calls are particularly likely to attract the male's attention when he is in a state of tension after having just fought with other males. Table 14 shows that, following an agonistic encounter with other males, the harem male was significantly more likely to approach a cycling female than any other female of his unit if she called ($\chi_1^2 = 6.707$, $p < 0.01$); if the cycling female did *not* call, the male was no more likely to approach her than any other female ($\chi_1^2 = 2.939$, $p > 0.05$). The male invariably inspected the oestrous female under these conditions, and in

Table 13
Frequency distribution of contexts in which oestrous females gave solicitation calls.

Context	Frequency	%
No obvious stimulus	326	39.1
Sexual interaction with male	75	9.0
Female looks at male while feeding	69	8.3
Female approaches male	50	6.0
Male approaches female	34	4.1
Female moves (no approach to male)	90	10.8
Male moves (no approach to female)	120	14.4
Female grooming male	23	2.8
Male in agonistic encounter elsewhere	34	4.1
Other behavior by male	10	1.2
Miscellaneous	2	<0.1

Source of data: 7 cycling females from 4 reproductive units sampled in 1974–75.

Table 14
Frequencies with which the harem male approached a cycling female after an encounter with an all-male group in relation to whether or not she gave a solicitation call.

| | Male Approaches: | | | |
	Oestrous female	Anoestrous female	Total	χ_1^2
Oestrous female calls	10	4	14	6.707
Oestrous female silent	15	13	28	2.939
Availability (%)	37.9	62.1		

Source of data: 7 cycling females in 4 reproductive units sampled in 1974–75.

a high proportion of cases he also mounted her. On average, females called on 14.7% of the occasions when their male was involved in encounters of this kind during the receptive part of the oestrous cycle, but on only about 2% of occasions during the period around menstruation (Table 15), a difference that is statistically significant ($\chi_1^2 = 14.545$, $p < 0.001$). The male was likely to approach the cycling female after

Table 15

Percentage of occasions when the male was involved in an encounter with an all-male group in which a cycling female gave a solicitation call and the male approached and mounted the cycling female.

	Stage of Oestrous Cycle[a]				χ_1^2 (B=D)	p
	A	B	C	D		
Female calls (%)	8.3	17.5	13.6	1.8	14.545	<0.001
Male approach/mount (%)	8.3	6.1	11.1	2.7	0.915	>0.10
Number of encounters	36	114	81	113		

Source of data: as for Table 14.

[a] Cycle stages are defined by Dunbar (1978b, p. 168): ovulation occurs during stage B, menstruation during stage D.

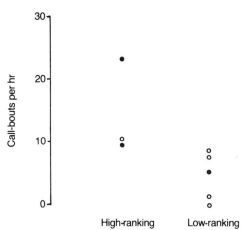

Figure 27   Mean frequencies per hour with which high-ranking (ranks 1–2) and low-ranking (ranks 3–5) females gave solicitation calls while in oestrus. Females are also classified by age into young (open circles) and old (filled circles). Source of data: as for Fig. 26.

an encounter roughly in proportion to the frequency with which she called.

Is the frequency with which a female called related to her status? Figure 27 gives the mean frequencies with which 8 females called during peak oestrus. Low-ranking females called less frequently than high-ranking females (Mann-Whitney test, $p = 0.056$ 2-tailed), but there was no obvious difference in relation to age.

This is surprising: we might expect high-ranking females to resort less frequently to tactical options, since they are already doing as well as they can. One possible explanation for the failure of this prediction is the presence of higher-ranking females who are likely either to interfere in the sexual interactions of subordinate females or to harass them (see Dunbar 1980b, Table 7). This may cause a subordinate female to suppress her vocal signaling in order to minimize the extent to which her condition attracts the attention of these females.

Do subordinate females succeed in getting the male to initiate a higher proportion of copulations? Apparently not: the male initiated 12 out of 37 copulations (32.4%) with 6 high-ranking (ranks 1–2) females and 13 out of 35 (37.1%) with 8 low-ranking (ranks 3–5) females. The difference is not statistically significant (2×2 $\chi_1^2 = 0.157, p > 0.60$). Nor are there any significant differences between low-ranking and high-ranking females in the rates with which the male mounted or ejaculated (Dunbar 1980b, Fig. 8).

Reproductive Tactics

Females can respond tactically in two ways at the physiological level. One is to step up their reproductive rates by reducing the inter-birth interval; the other is to manipulate the sex ratio of the offspring they conceive to take maximum advantage of the different reproductive potentials of the two sexes.

REDUCING THE INTER-BIRTH INTERVAL

Gelada have an unusually long inter-birth interval compared with most baboons and macaques. Since most of this is accounted for by a prolonged post-partum amenorrhea, females could increase their birth rates by coming back into oestrus earlier. That they can in principle do this is shown by the fact that, following the takeover of a reproductive unit, many of the unit's females come into oestrus prematurely (Mori and Dunbar 1984); as a result, the mean inter-birth interval for these females is reduced from the normal 2.14 years to 1.64 years (see Dunbar 1980a, Fig. 6). There is, however, no indication that low-ranking females have significantly shorter inter-birth intervals than high-ranking females (Fig. 28: $r_s = -0.042, n = 28, p = 0.416$ 1-tailed).

Theoretical considerations suggest that reducing the inter-birth interval will usually be possible only at some cost in terms of infant or maternal survival (Altmann 1980). Can we identify any such costs that might militate against gelada females pursuing this option regularly?

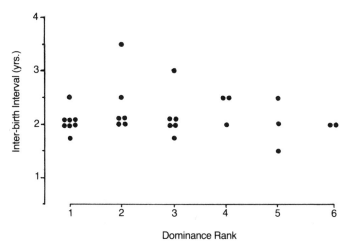

Figure 28   Inter-birth intervals plotted against dominance ranks for 51 females from 11 units sampled in 1974–75. Intervals were estimated either from the age of the previous infant for females who gave birth during the study (11 females) or by adding the length of the gestation period (6 months) to the age of the female's youngest offspring at the time she started to cycle again (17 females).

Although offspring have been observed suckling as late as 2 years of age, the amount of milk produced after the infant is about 12 months old is probably rather small. It seems unlikely, therefore, that premature cycling by the mother would have a significant impact on the previous infant's chances of survival. None of the 5 infants aged 6–12 months whose mothers came back into oestrus following a takeover showed any signs of increased risk of mortality, at least during the subsequent 3–4 months.

The explanation might lie not so much in the previous infant's chances of survival as in the new infant's prospects. Most infants are born in November–March during the dry season, having been conceived with the main period of grass growth following the previous rains (Dunbar 1980a). Approximately 40% of infants, however, are born in a secondary birth peak in May–August (just prior to and during the rains), having been conceived with a minor flush of grass during December–January. An infant born during this second peak faces 4–5 months of severe weather in its porous natal coat. Neonatal mortality is closely correlated with the number of wet season months the infant has to live through in its natal coat (Fig. 29). Conceiving at 18-month intervals would increase the number of infants lost through increased neonatal

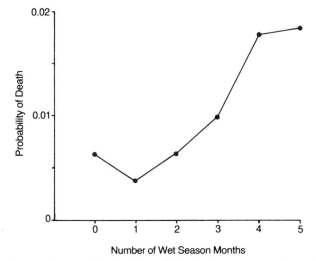

Figure 29   Monthly probability of death in relation to the number of wet season months that an infant spent in its porous natal coat. Based on a 3-point
running mean of a total sample of 546 infant-months of data (summarized in
Dunbar 1980a, Table 5).

mortality during the wet season. Nonetheless, the increment in risk is
not so severe that, other things being equal, a female could not expect
to do better over a reproductive lifetime by doing this. With an expectation of life of $e_4 = 9.8$ years at reproductive maturity and a mean
inter-birth interval of 2.14 years, a female can expect to produce 4.57
offspring in her lifetime. By reducing the inter-birth interval by 6 months,
she could produce 5.97 offspring, half of whom would be born during
the wet season and half during the dry season. Of the 2.98 infants born
during the wet season, 0.06 would, on average, die, leaving a net gain
at the end of the mother's lifetime of 1.34 offspring (29.3%). This is a
considerable gain, so why don't females pursue this option?

The explanation might lie in the female's own survival. Adult mortality was also correlated with the seasonal distribution of rainfall (Dunbar
1980a), and the strain on the female of lactating during a period of
severe climatic stress may be too great to warrant the risk. A female
at the midpoint of her reproductive lifespan with $e_{8.5} = 4.9$ years would
have to die 3.61 years early to offset a gain of 1.34 offspring at the
observed birth rate for females aged 8–13 years (0.368 births per year:
see Fig. 11). This would require an increase in the mortality rate of the
order of 380%. Although the available data are limited, there was no
suggestion that the mortality rate for lactating females was significantly

higher during the wet season than that for other females. Nonetheless, the altitudinal cline in birth rates (see Ohsawa and Dunbar 1984) implies that the females at Sankaber are doing about as well as they can. This suggests that the limitation may lie not so much in post-natal mortality rates as in whether or not the female is physiologically capable of conceiving again so soon on a regular basis (see also Altmann 1980).

MANIPULATING THE SEX RATIO OF OFFSPRING

The other reproductive tactic open to a female is to adjust the sex ratio of her offspring so as to capitalize on the differential reproductive abilities of the two sexes.

Generally speaking, the variance in male reproductive output is greater than that for females. Consequently, as Trivers and Willard (1973) have pointed out, females might be expected to have more or fewer male offspring in proportion to the risk they can afford to take in terms of their own personal reproductive outputs. In this way, a female can influence her contribution to the species' *future* gene pool by contributing a higher proportion of *genes* (or at least as many) as does a competitor that produces as many (if not more) actual offspring.

Since high-ranking females produce more offspring, they can afford to risk losing some and should, therefore, tend to have a higher proportion of males than do lower-ranking females. By the same token, females should also produce more sons when young and more daughters when old, since young females can make up any losses at a later date. An old female with enough time left for only one more offspring to come should play safe and produce a daughter.*

Figure 30 gives the proportion of males among the offspring of females of different dominance ranks, based on the number of offspring living at the time of the census. There is no obvious correlation between the two variables ($r_s = 0.059$, $p = 0.890$ 1-tailed). These data, however, suffer from the same problem as beset the analysis of rank-dependent birth rates given in Dunbar (1980b): the offspring of any one female derive from an extended period of time during which the female's rank could well have undergone radical changes. The actual

---

* This is an explicitly genetic argument, and it is important to note that an individual's relative contribution to the gene pool depends on the demographic state of the population (see Schaffer 1974, Charlesworth 1980). In expanding populations, offspring produced while the female is young will be worth more than offspring produced late in life, while the converse will be true in declining populations. Since the gelada populations are expanding (Dunbar 1978a, 1980a), females *should* produce sons when young and daughters when old.

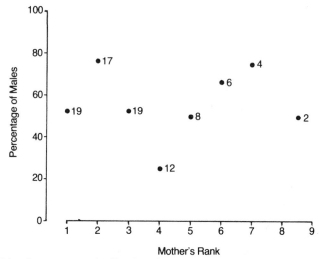

Figure 30   Percentage of offspring that were male plotted against mother's dominance rank. Sample sizes are indicated beside each point on the graph. Source of data: 51 females from 11 units sampled in 1974–75.

number of births during the study was too small to analyze in detail; however, pooling data for different ranks does not suggest that there was a clear difference in the ratio of male:female infants born to high-ranking (ranks 1–2) and low-ranking (ranks 3–10) females (6:1 and 5:3, respectively). Nor was there any clear tendency for primiparous mothers to produce males more often than multiparous mothers (4:1 vs. 7:3 respectively), though the sample size is again too small for the difference to be statistically meaningful. Table 16 gives the frequencies of male and female offspring less than 4 years old for females of different age classes. None of the distributions differs from an even sex ratio, nor does that for first-born offspring ($\chi^2_1 < 0.862, p > 0.30$).

Clutton-Brock and Albon (1982) have pointed out that the variance in sex ratios at birth is often very fine, even in those cases where there is a consistent bias in favor of one sex; consequently, large samples are invariably required to demonstrate the presence of any differences statistically. That no significant differences were detected in the present case does not necessarily mean that such effects do not exist. Indeed, the data for gelada might be taken to suggest that high-ranking females do produce a slightly higher proportion of males than low-ranking females, as Trivers and Willard (1973) predict. If so, then these results are contrary to those reported by Altmann (1980) for free-living *Papio cynocephalus* and by Simpson et al. (1982) for captive *Macaca mulatta*.

Table 16
Sex ratio of all offspring less than 4.25 years old for females of various age classes, together with the sex distribution of the first-born offspring.

| Sex of Offspring | | Female Age Class | | First-born Only[a] |
	Subadult	Young adult	Old adult	
Male	11	15	24	12
Female	11	14	21	17

Source of data: 13 reproductive units sampled in 1974–75.
[a] Subadult and young adult females only.

These studies showed clear tendencies for high-ranking females to produce more *female* offspring than low-ranking females (see Clutton-Brock 1982). However, we are not entitled to reject the Trivers and Willard hypothesis out of hand unless we can be certain that there are no other proximate factors influencing the situation in such a way that the hypothesis itself would make different predictions in the two cases. Among these will be the nutritional condition of the females and the access to high quality food sources that high rank confers upon them (see Clutton-Brock 1982, Clutton-Brock and Albon 1982). Social considerations may also be important. The analyses of Chapter 7, for example, suggested that a first-born daughter may be particularly important to a female in terms of coalition formation and its effects on her lifetime reproductive output. It may therefore be significant that there is a slight preponderance of females among first-born offspring (Table 16). On the other hand, in species such as baboons and macaques where coalition size may be an important determinant of a female's rank, females who consistently produce a high proportion of female offspring may achieve higher rank than females whose offspring are male-biased.

Parental Investment Tactics

A female might be able to compensate for a poor birth rate by investing more in each offspring that she produces, thereby giving each one a better chance of reaching reproductive age and/or of competing successfully in the reproductive arena. Data to test this prediction are available from a study of mother-infant relationships carried out at Sankaber by Patsy Dunbar during 1974–75. In all, 10 infants were studied from birth over periods of 4–9 months. During each sample period (weeks 1, 2, 3–4 and monthly thereafter), an average of 3.6 ± 0.88

hrs of detailed data were obtained for each mother-infant dyad. In each case, the activities and spatial relationships of both mother and infant were determined at 30-sec intervals from an instantaneous scan, while interactions were recorded on a continuous basis.

The behavior most likely to be directly relevant to parental investment is the time that the infant spends on the nipple. Not all the time an infant is on the nipple is spent suckling (i.e. receiving direct investment through lactation): as with all primates, gelada infants spend significant quantities of time on the nipple for reasons of comfort and "security" (as well as for reasons of transport, especially during the first month when the infant is carried clinging to the mother's ventrum). However, the infant's presence on the nipple becomes an increasingly serious problem for the mother as the infant grows because it interferes with her ability to harvest grass blades in a seated position. Females clearly experience a great deal of inconvenience due to disrupted time budgets and reduced feeding opportunities from the infant's persistent attempts to gain access to the nipple at inappropriate times. This results in increased frequencies of so-called "weaning" behavior on the part of the mother and "weaning tantrums" by the infant (these clearly having less to do with the quantity of suckling then with the timing of suckling).

Figure 31 plots the percentage of time that infants of 5 high-ranking females (ranks 1–2) and 5 low-ranking females (ranks 3–6) spent on the nipple in each of 11 successive sample periods. The infants in the two categories were balanced for sex, mother's parity, and harem size (the first two both being factors that significantly influence most of the behavioral variables that were sampled). Overall, the infants of low-ranking females spent significantly more time on the nipple than those of high-ranking females (Wilcoxon matched-pairs test comparing median values for each sample period, $z = 1.689$, $p = 0.046$ 1-tailed), although in only one period (month 6, the month when infants start to feed significantly for themselves) was the difference in individual scores significant (Mann-Whitney test, $p = 0.016$ 1-tailed). To the extent that the mother controls the infant's access to the nipple, this suggests that low-ranking females are slightly more tolerant than high-ranking mothers, thereby investing more time (and perhaps energy) in their offspring as a result.

The cost of nipple-contact should be reflected in a disrupted time budget for the mother. Moreover, lactation is an extremely costly activity for female mammals, it being a relatively inefficient use of energy (see Portman 1970, Altmann 1980): in the gelada, lactating females spend approximately 30% more time feeding each day than non-lactat-

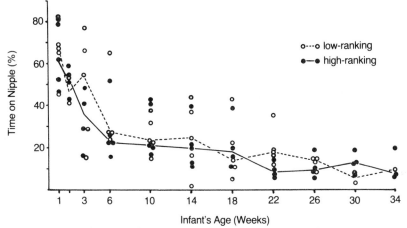

Figure 31   Percentage of time that infants spent on the nipple during their first 9 months. Infants are distinguished according to the mother's dominance rank into high-ranking (ranks 1–2) and low-ranking (ranks 3–6). The median values are indicated for each group. Source of data: 10 mother-infant dyads sampled during 1974–75.

ing females (Dunbar 1983d, Fig. 2). Figure 32 shows that, overall, low-ranking females spent significantly less time feeding than high-ranking females (Wilcoxon test comparing median values for each period, $p = 0.020$ 2-tailed), the individual scores being significantly different in months 6 and 7 (Mann-Whitney tests, $p < 0.05$ 2-tailed). Note how the time spent feeding increases steadily for both groups of females as the growing infant's energy demands increase with age, to reach a peak in month 6. Significantly, feeding time declines thereafter for high-ranking females as the infant begins to feed for itself, whereas low-ranking females seem to maintain a high feeding rate for rather longer. These results seem to indicate that low-ranking females are building up an energy debt as a result of spending less time feeding than they ought to in the early months, this deficit (presumably reflected in a marked loss in weight) being offset by continuing to feed at a high level after the peak in the infant's energy demands. (There seems to be no good reason to suppose that the infants of low-ranking females require more suckling than those of high-ranking females, nor any reason to suppose, at least in the case of the gelada, that low-ranking females are prevented from spending more time feeding by more dominant individuals.)

Taken together, these data suggest that low-ranking females invest

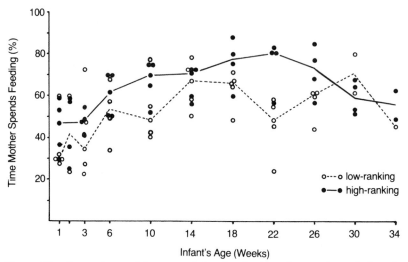

Figure 32   Percentage of time spent feeding by lactating mothers during the first 9 months post-partum. Legend and source of data as for Fig. 31.

proportionately more heavily in their offspring than do high-ranking females and that they do so at greater potential risk to their own survival. However, because of the inherent complexity of developmental processes, it is not entirely clear just what effect this investment has on the infant's long-term reproductive prospects. All we know for certain is that low-ranking mothers bear an increased cost (equivalent on average to about 5% more time spent on nipple overall), but we cannot say whether or how this cost might offset their lower basal output rates in the long run.

How Important Are Tactics?

There is little evidence to suggest that females consistently use any of the tactical options other than parental investment to offset the reproductive disadvantages of low rank. There are two likely reasons for this. First, all the tactical options are just as advantageous for high-ranking as for low-ranking females. Secondly, none of them provides a sufficiently large pay-off (either because the gain is too small or the cost too high) to provide a useful antidote to loss of reproductive expectations at a strategic level. Moreover, the range of options open to a female at a strategic level is clearly considerable, and probably provides much more flexibility over the course of a lifetime than she can gain from the relatively small benefits generated by any of the tactical options.

10 The Male's Loyalty Problem

The female's problems relate strictly to intra-unit competition among females, with matrilineal coalitions forming a crucial basis from which she operates. The male, on the other hand, is effectively on his own and his problems are twofold. First, in order to breed at all, he has to gain control over a harem of females. Secondly, having gained a harem, he has to prevent other males from taking it away from him. His first task is not easy, and his second is compounded by the females' own strategic designs.

In Chapter 7, we saw that females are liable to desert their male in favor of another and it was suggested that poor reproductive performance was the main factor precipitating desertion. This problem reaches crisis proportions for the male when desertion occurs in favor of a male who is challenging his hegemony over the harem, for if one female deserts, then it is virtually certain that all his other females will follow suit in fairly short order (see Dunbar and Dunbar 1975). Moreover, once a male has lost his harem, he gets no further opportunities to own another. (At least, we never saw any instances where a former harem-holder reacquired control over a harem or, indeed, even tried to do so.) Instead, defeated harem males remain as elderly followers in their own former units, though without the benefits of sexual access to any of the females. The reasons for this will be discussed in Chapter 15. For the moment, it is sufficient to note that a male gets one, and only one, opportunity to own a breeding harem in his lifetime.

In his experimental studies of gelada reproductive units, Kummer (1975) postulated that a female's "loyalty" to her male (which we can define as the probability that she will not desert him for another male over a given period of time) was proportional to the frequency with which she groomed with him. If this hypothesis is correct, then a harem male is likely to face serious trouble as the size of his harem increases. This is because, while each female will presumably need to be groomed for some specifiable amount of time if her loyalty is to be guaranteed, the amount of social time he has available is limited. Sooner or later, the male will encounter a problem in time-sharing.

In this chapter, I first show how increases in harem size affect group structure and then how this affects female loyalty. I go on to examine the problem this creates for the male and conclude by considering some ways in which males might seek to minimize the likelihood of being deserted by their females. The analyses are based on 14 reproductive units whose social structure was sampled during the 1974–75 study and 11 units sampled during the 1971–72 study. The social structure of each unit was determined in exactly the same way, using the scan-sampling technique described in Chapter 1. It should be noted that sampling was carried out only while at least one adult member of the unit was interacting with another member of the unit (where by "interacting" I mean positive, friendly interactions: 98.6% of 2881 scans of interacting dyads recorded from 3 randomly chosen units involved grooming). The time base used in the analyses of these data is termed *potential social time;* no individual spends all of this time interacting. Frequencies of social interaction are given as percentages of potential social time, not of total daytime or time spent actually interacting. The total amount of time actually spent interacting by an individual is referred to as its *net social time.*

On average, 218 scan samples (range 33–681) were obtained for each unit, with large units generally being sampled for longer than small ones. As few as 25–50 scans were sufficient to describe the social relationships among a unit's members correctly. (For further details, see Dunbar 1983b,c.)

Demographic Effects on Group Structure

Reproductive females do not necessarily interact with all members of their units; indeed, in large units, many pairs of females may never interact (Dunbar 1983b). This is mainly due to the fact that females prefer to groom intensively with their close female relatives rather than attempt to interact with all members of the unit.

To determine the consequences of this preference on the structure of the unit, social relationships within units (as determined from the scan samples) were analyzed in relation to unit size. To assess the degree of social cohesion in reproductive units, I have used three different indices that measure the degree of clustering within their social networks: (1) the minimum discriminant required to generate at least two subgroups; (2) the numbers of non-interacting subgroups or cliques (*sensu* Sade 1972) that are generated by a discriminant (or cut-off) equivalent to 10% of time spent interacting; and (3) the mean diversity in the choice of social partners by the adult members of the unit.

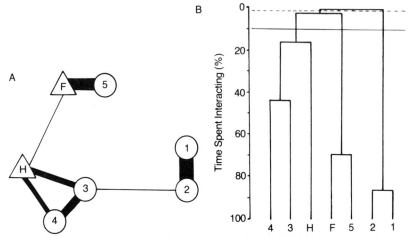

Figure 33 (a) Sociogram of the 1971–72 unit, H39: H indicates the harem-holder, F the young follower. (b) Tree-dendrogram of the relationships among the adults of the same unit, obtained from a single-link cluster analysis. The solid line parallel to the x-axis is the level used to generate 10%-subgroups; the broken line is the minimum discriminant that will split the unit into at least two non-interacting subgroups.

The first two indices derive directly from the interaction matrices for each unit. If these matrices are transformed into "single link" dendrograms (see Morgan et al. 1977) as shown in Figure 33 for the 1971–72 unit H39, indices (1) and (2) can be read off directly. These are indicated in Figure 33b by the two horizontal lines drawn across the dendrogram. The first measure corresponds to the highest horizontal line drawn across the dendrogram (the dashed line parallel to the x-axis) that would split the group into two separate clusters. On a sociogram, it would correspond to the lowest frequency of interaction that would, if all less frequent interactions were removed, result in the unit appearing as two separate subsets of individuals between which there were no interactions. In the case of the unit illustrated in Figure 33, this corresponds to a value of 2% of time spent interacting. The second index measures the number of interacting subgroups in the unit in relation to unit size. A discriminant of 10% was used, since this value was found to be a useful level when considering male and female social relationships (see Dunbar 1979b, 1983b,c) and receives some retrospective empirical support from Figure 34 (see below). A 10%-subgroup was defined as the number of individuals who were included in the same cluster when a discriminant of 10% was applied to a dendrogram.

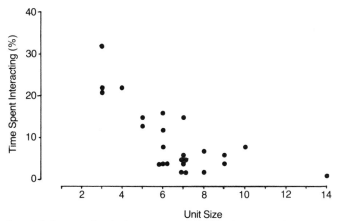

Figure 34   Minimum discriminant on a tree dendrogram of intra-unit social relationships (see Fig. 33) required to generate at least two non-interacting subgroups, plotted against the number of adults in the unit. Source of data: 11 reproductive units sampled in 1971–72 and 14 units sampled in 1974–75. (Redrawn from Dunbar 1983e)

In the case of the unit illustrated in Figure 33b, applying a discriminant at the 10% level (the solid line parallel to the x-axis) yields 3 subgroups.

Figure 34 shows that as unit size increases, the value of the subgrouping discriminant declines and appears to reach an asymptote at a value between 5–10% at a unit size of 6–7 adults. A regression fitted by the method of least squares to a log/linear transformation of the data has a slope that is significantly less than zero ($r^2 = 0.503$, $t_{23} [b = 0] = -4.824$, $p \ll 0.001$).

The second index (the number of discrete 10%-subgroups or cliques) is plotted against the number of adults in the unit in Figure 35. This clearly increases with unit size (linear regression on log/linearly transformed data: $r^2 = 0.687$, $t_{23} [b = 0] = 7.098$, $p \ll 0.001$), again indicating increasing fragmentation as the number of potential interactees in the unit increases. In fact, the distribution of points on the graph suggests that there may be a threshold for fragmentation at a unit size of 6–7 adults.

In Figure 36 the sizes of all such subgroups are plotted against the number of adults in the unit. It is apparent that, initially, the whole unit is a closely integrated social entity, but that, as unit size increases beyond about 6 adults, it starts to fragment into small subgroups that do not often interact. Of particular significance in the present context is that subgroup size appears to decline as unit size increases beyond

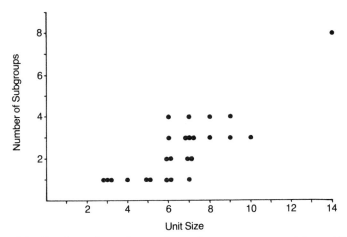

Figure 35   Number of non-interacting subgroups generated by a 10% fre-
quency-of-interaction discriminant (see Fig. 33), plotted against the number of
adults in the unit. Sample population as for Fig. 34.

this level: mean subgroup size declines from 3.83 in small units (3–5
adults) to 2.61 in intermediate units (6–7 adults) and 2.32 in large units
(8–14 adults). The distribution of subgroup sizes is significantly different across unit sizes ($\chi_2^2 = 12.007$, $p = 0.002$), although partitioning
degrees of freedom (following Goodman 1968) reveals that only the
difference between small and larger units is significant (6–7 vs. 8–14
adults, $\chi_1^2 = 1.325$, $p = 0.20$; 3–5 vs. 6–14 adults, $\chi_1^2 = 10.682$, $p = 0.001$).

The third index is a measure of the extent to which the members of
a unit divide their social time evenly among themselves. It is the average of the diversity of each adult's social interactions with the other
adults in its unit. Each individual's index was determined from the
interaction matrices using Shannon's heterogeneity index,

$$H = -\sum_i p_i \ln p_i,$$

where p_i is the proportion of net social time devoted to individual i.
This was adjusted to compensate for unit size to give Buzas and Gibson's (1969) evenness index,

$$H^* = \exp(H)/n,$$

where n is the number of interactees in the sample (see Peet 1974).
This statistic ranges between $H^* = 0$ (when the subject interacts with
only one individual) and $H^* = 1$ (when its net social time is evenly

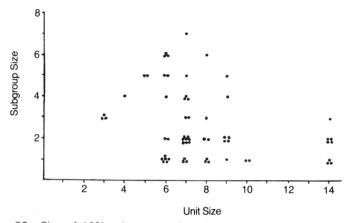

Figure 36   Size of 10%-subgroups, plotted against the number of adults in each unit, for the data given in Fig. 35.

divided among all the other members of the unit). The mean of the individual values of H^* was then determined for each unit. In the case of unit H39 (Fig. 33), the mean value was $\bar{x}(H^*) = 0.1964$.

The values of $\bar{x}(H^*)$ for each of the 25 units in the sample are plotted against unit size in Figure 37. It is evident that, as a unit's size increases, so its members distributed their social time less evenly. A regression fitted by least squares to a log/linear transformation of the data was again significant ($r^2 = 0.607$, t_{23} [$b = 0$] $= -5.964$, $p \ll 0.001$).

We can check that these are genuine dynamic changes by determining what happens to these indices when one member of a unit dies. The structure of social relationships before and after a death was sampled in 4 units. In addition, data are available in one case where unit size increased due to the immigration of a subadult female (unit H22 in 1972). (Changes in unit size due to takeovers and follower-entry have been discounted for this purpose in view of the highly disruptive consequences of these events.) In all 5 cases, the three indices changed in the predicted direction (binomial tests: $p = 0.031$ 1-tailed for each index). Figure 38 illustrates this with the results for the diversity of interactions index.

Kawai and Mori (1979, Fig. 8.1) obtained analogous results for the Gich population. They found that the mean distance both between all adults and between the harem male and his females increased with unit size. They interpreted this to imply that large units are less cohesive than small units.

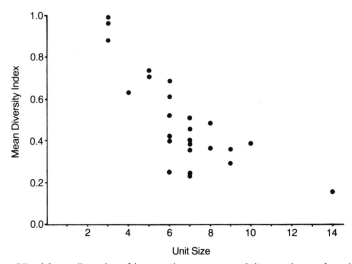

Figure 37 Mean diversity of interactions among adult members of each unit, measured by the Buzas–Gibson index, H^*, plotted against the number of adults in the unit. Sample population as for Fig. 34.

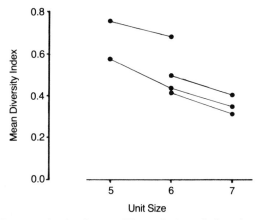

Figure 38 Changes in the Buzas–Gibson index of diversity of interactions, H^*, before and after a death for 4 units sampled in 1975 and the immigration of a female for one unit sampled in 1972.

This fragmentation of the unit can be attributed partly to the female's failure to interact with anyone outside her own immediate matriline and partly to the male's inability to link the female subgroups together by grooming with all of them at the requisite level. The latter

Table 17

Frequencies with which small and large units were taken over relative to their frequencies in the population.

	Number of Females in Unit	
	1–4	5–10
Number of units taken over	0	7
Number of units in population	25	23

Source of data: 31 units sampled in 1971–72 and 17 units sampled in 1974–75.

factor, of course, is itself in part a consequence of the females' lack of interest in grooming with the male.

Consequences of Social Fragmentation

That the male's predicament is serious can be shown by three sets of data. First, takeovers of units by males from all-male groups were not randomly distributed. A total of 7 takeovers were observed during the two field studies: a comparison of the sizes of unit taken over with those available in the sample population shows that units with less than the median number of females (4 reproductive females) were significantly less likely to be taken over than larger units (Table 17: binomial test, $p = 0.012$). Second, Figure 39 shows that, for units with more than 4 reproductive females, males who were subsequently taken over had more females in their units with whom they did not interact than males who were not taken over, though the difference fails to reach statistical significance (Mann-Whitney test, $z = 1.687$, $p = 0.092$ 2-tailed). Third, Figure 40 shows that males who were subsequently taken over spent significantly less time grooming with each of their females on average than did males who were not taken over (means of 5.8% and 9.8%, respectively; Mann-Whitney test, $p = 0.037$ 2-tailed).

The last two differences cannot be attributed to differences in harem size because the distributions of harem size were identical. Nor can they be attributed to differences in the ages of the males (see Dunbar 1983c), as all the males except one were classed as mid-prime in age. It is no doubt significant that, in Figure 40, the overall average for males who were not taken over is above the critical limit required to generate two subgroups (see Fig. 34), whereas that for males who were taken over is well below this value. Indeed, males who were later taken over were below this value (taken as 7.5%) significantly more often

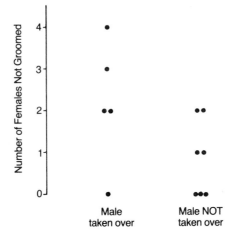

Figure 39   Number of females in his unit with whom the harem male did *not* interact, in relation to whether or not he was later taken over, for males with large units (i.e. more than 4 reproductive females) who were at risk of being taken over. Source of data: 12 units sampled in 1971–72 and 1974–75.

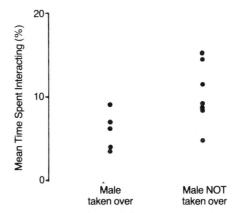

Figure 40   Mean percentage of a male's potential social time that he spent interacting with each of his females in relation to whether or not he was later taken over for males with large units (more than 4 females). Sample population as for Fig. 39.

than males who were not taken over (Fisher exact test, $p = 0.05$). These effects are independent of the tendency for older males to be less diligent in servicing their relationships with their females (see Dunbar 1983c). The relationship with age is an additional contributory factor, but its importance is relatively small compared with the influence of harem size.

A Problem in Time-Sharing

To summarize so far: the male's problem is that, as the size of his harem increases, his available social time becomes insufficient to groom each female often enough to guarantee her loyalty, and, apparently as a consequence of this, he becomes increasingly likely to lose his harem. Since displacement as a unit leader effectively signals the end of his breeding career, we would expect males to have evolved strategies to counteract these effects in order to delay takeover for as long as possible.

The male would seem to have only three options: (1) he could increase the amount of time he spent interacting; or, failing this, he could either (2) continue to distribute his time evenly among all the females, even though he groomed each one for less than the critical limit, or (3) distribute his time evenly until he reached the loyalty limit and then ignore the presence of any additional females. Although the *average* loyalty of the females would be the same in the last two cases, the third strategy has the advantage that the loyalty of at least some of the females is assured. This may be sufficient to prevent the remaining females from deserting if desertion is an all-or-none phenomenon (see Chapter 11).

The first possibility seems intuitively unlikely, since the male's time budget is not infinitely flexible. Gelada are bulk feeders on a low-quality food source and need to spend 50–75% of their time foraging in order to obtain their daily nutritional requirements (Iwamoto and Dunbar 1983). At Sankaber, a male could spend at most 34.3% of his day in social activity (see Table 3). Since he already spends 20.5% of his time interacting, he could increase his social time by at most 67%. Thus, even if he spent all his available time interacting, he would not be able to groom more than about 8 females at the requisite level. Moreover, the time that a male can devote to grooming his females is constrained by the amount of time which his females are prepared to spend grooming with him. Unless the females are willing to interact, increasing the amount of time during which *he* is willing to interact will avail him nothing. In practice, this seems to be the case, for the amount of time a male spends interacting does not increase with harem size: if anything, it decreases with increasing harem size (Fig. 41: linear regres-

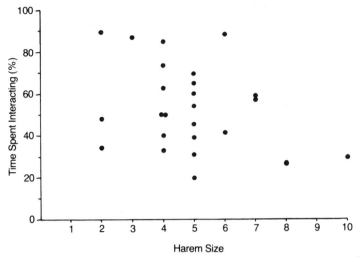

Figure 41   Percentage of potential social time that harem males spent interacting with the females of their units, plotted against harem size. Sample population as for Fig. 34.

sion fitted by least squares to log/linearly transformed data, $r^2 = 0.101$; $t_{23} = -1.611$, $p = 0.934$ 1-tailed).

We can test the second possibility by examining the change in the time that the male spent interacting with his main female partner in relation to harem size (Fig. 42). A linear regression (fitted to log-linearly transformed data by the method of least squares) gives a slope that is significantly less than zero ($r^2 = 0.211$; $t_{23} = -2.478$, $p = 0.002$ 1-tailed). It is evident from the distribution of points on the graph that the relationship between the two variables is not strictly linear, for time spent interacting seems to reach an asymptote at a value of about 10%. If males did groom all their females equally, we would expect the points for the larger harem sizes to fall well below 10% since the male has only 54% of potential social time available to distribute among his females (Fig. 41). Figure 43 shows that all the values for time spent grooming the partner fall well above the main diagonal when plotted against the mean value expected if the male distributed his net social time evenly among his females. A least-squares linear regression fitted to these points on a double-log scale has a slope that is greater than 0 ($r^2 = 0.522$; $t_{23} [b = 0] = 5.011$, $p < 0.001$) and a y-intercept that is greater than 0 ($t_{23} [a = 0] = 3.245$, $p = 0.004$): males devoted significantly more time to their partner females than they ought to have done had they tried to groom their females evenly.

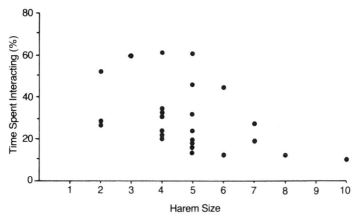

Figure 42   Percentage of potential social time that harem males spent interacting with their main female grooming partners, plotted against harem size. Sample population as for Fig. 34.

The third hypothesis can be tested by plotting the number of females that the male interacted with against harem size (Fig. 44). This also permits an alternative test of the second hypothesis, since, if the male had tried to groom with all his females, the points would lie on the main diagonal. Initially, the male does interact with each of his females, but the number with whom he interacts appears to reach asymptote at a harem size of 5–6 females. A least-squares linear regression (fitted to the data on a double-log scale) has a slope that is significantly less than 1 ($r^2 = 0.521$; t_{23} [$b = 1$] $= -2.462$, $p = 0.022$). (The slope is also significantly greater than 0: t_{23} [$b = 0$] $= 5.001$, $p < 0.001$.) Thus, males seem to pursue a two-stage strategy in which they attempt to groom with all their females at first, but cease doing so once the amount of time they can spend grooming with each of them reaches a critical lower limit.

Risk-Minimizing Substrategies

If the male ignores any new females that are added to his harem, these females will always be high desertion risks whenever another male attempts to take the unit over. Because females seem to make a collective decision about allegiance to their males, desertion by any one of these new members might precipitate a general desertion by the other females. However, for the reasons discussed in Chapter 7, the proba-

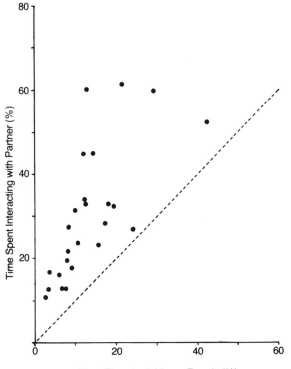

Figure 43 Percentage of potential social time that harem males spent inter-acting with their main female grooming partners, plotted against the mean time each male had available to interact with all of his females. The broken line indicates the main diagonal around which all the points would cluster if males groomed all their females equally. Sample population as for Fig. 34.

bility of desertion is unlikely to be constant for all females, but rather will vary with a female's age and dominance rank. This offers the male scope for selectively grooming his females in order to maximize overall loyalty while minimizing his costs in terms of time spent interacting.

Four alternative substrategies suggest themselves: (1) the male could groom low-ranking females more than high-ranking ones, because the risk of desertion is higher for low-ranking females; (2) the male could groom young females more than older females, for the same reason; (3) the male could groom just one member of each female grooming dyad, relying on the strong female-female bonds to guarantee the other member's loyalty; (4) in a more sophisticated version of the last pos-sibility, the male could groom whichever member of the female dyad

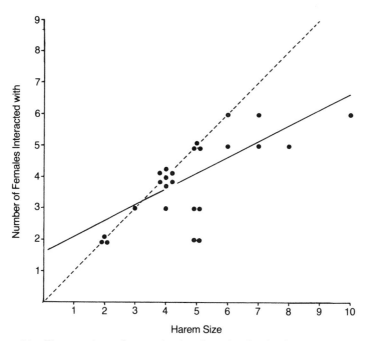

Figure 44   The number of reproductive females in the harem with whom each male interacted, plotted against harem size. The broken line indicates the main diagonal along which all the points would lie if males groomed all their females equally; the solid line gives the least squares linear regression fitted to the data. Sample population as for Fig. 34.

had the higher desertion risk. Because dyad members occupy adjacent ranks (Dunbar 1980b), dominance is unlikely to be a relevant basis for choice in this last respect; however, the two females are invariably of different ages (Dunbar 1979b) and this might generate a sufficiently large loyalty differential between the two to make grooming with the younger female a practical proposition.

The first two hypotheses can be tested together by examining the frequencies with which males groomed subordinate and young adult females in those units where the male did not groom with all his females. This also allows us to look for an interaction effect between these two variables. The relevant data were examined in two ways. First, only units where the male was observed not to interact with at least one female were considered. Since this yielded a sample of 23 females based on only 4 units, the criterion was raised from 0% to 5% of time spent interacting and the data reanalyzed to determine which

Table 18

Number of females with whom the male spent more than 5% of his time grooming in units in which he did not interact with all the females at this level, with expected values (given in parentheses) calculated on the assumption that the male selected his grooming partners at random amongst the females in his unit irrespective of their age or dominance status.

| Female | Number of Female Interactees | | |
Dominance Ranks	Young females	Old females	Total
1–3	4 (4.9)	10 (7.8)	14
4–10	2 (4.9)	10 (8.3)	12
Total	6	20	26

Source of data: 7 units sampled in 1974–75.

females the male interacted with for more than 5% of his time. This increased the sample size to 7 units, though it only increased the number of female interactees to 26. Table 18 gives the latter set of data, together with expected frequencies assuming a random choice of partners among the females of these classes in proportion to their availability in the sample population. With $\chi_3^2 = 2.850$ ($p = 0.415$), it is apparent that the males made no attempt to interact disproportionately often with either young or subordinate females. The same result is obtained if the stricter criterion of 0% time interacting is used ($\chi_3^2 = 0.710$, $p = 0.871$). (There are three degrees of freedom in both cases for the 2 × 2 contingency table because expected values are not calculated from the marginal totals but from probabilities that are independent of the data in the table; consequently, since only one parameter, the grand total, is estimated from the data, only one degree of freedom is lost: see Goodman 1968, Maxwell 1968.)

The third possibility was that the male might groom with only one member of each dyad. Table 19 gives the numbers of dyads (from 9 units in which the male did not groom with at least one female) in which the male groomed with zero, one, or both members. An expected distribution is also given, calculated on the assumption that the male groomed at random with his females, irrespective of their dyad membership. The probability that the male would groom with any one female was taken as the observed proportion of dyad members with whom he actually groomed (i.e. 0.630). With $\chi_1^2 = 0.843$ ($p > 0.30$), it is clear that the male distributed his grooming at random. (In this case, two degrees of freedom are lost, since two parameters, the grand

Table 19

Number of members of female grooming dyads with whom a harem male groomed; expected values are calculated on the assumption that males groomed randomly with females irrespective of their dyad membership (with the probability of interaction with any female taken as the observed rate in the tabled sample).

Number	Number of Female Interactees			
of Dyads	0	1	2	Total
Observed	4	12	7	23
Expected	3.2	10.7	9.1	

Source of data: 3 reproductive units sampled in 1971–72 and 6 units sampled in 1974–75.

total and the probability of grooming with any one female, were estimated from the data in the table.)

The fourth possibility is, by definition, excluded by the results in Table 19. Nonetheless, we can check this by determining whether males are more likely to groom with the younger (i.e. potentially less loyal) of the two females in a dyad. Out of 16 dyads in 9 units in which the male did not interact with at least one of his females, he interacted most with the younger of the two in 6. If post-puberty juveniles are excluded on the grounds that males are sexually less interested in juvenile females than in adults (Dunbar and Dunbar 1975), the male interacted with the younger female in 5 out of 10 dyads. Neither distribution is significantly different from an expectation of 50:50 (binomial tests, $p \geq 0.227$ 1-tailed).

In conclusion, it seems that males make no special effort to maximize the average loyalty of their females by interacting more frequently with those females who are the greatest risks. This seems to be mainly a consequence of the fact that males are not free to groom with their females in the way they would wish.

Conclusions

The results presented in this chapter lend some support to Kummer's (1975) hypothesis that a gelada female's loyalty to her male is a function of the extent to which he grooms with her, at least insofar as the probability that a male will lose his females by mass desertion does seem to be a function of the extent to which he grooms with them. Successful takeover of a unit by a male from an all-male group occurs only in those units whose social structure has begun to show signs of fragmentation, because only in those units is the loyalty of at least some of the females sufficiently weak for them to be willing to transfer their allegiance to a new male.

The analyses suggest that the male makes only a partial attempt to forestall his ultimate fate: once his harem increases above a threshold size, he makes no effort to maximize the individual loyalty of all of his females. Since females desert their male on the basis of what amounts to a collective decision, simply ensuring that some females are less willing to desert him may be the best that he can do. Given the constraints on his time, it seems that the limit on this is about 5 females. It is likely that at least some of the substrategies that a male could pursue are probably not, strictly speaking, possible options. In particular, the suggestion that a male might groom with only one member of a grooming dyad probably expects too much, since females have time to groom extensively with only one individual, and that individual is preferentially a female relative (Dunbar 1983b).

Note the correlation between the median harem size in this population (4 reproductive females: see Fig. 5), the critical size at which social fragmentation sets in (approximately 5 females: see Figs. 33–37), the largest number of females that a harem male can interact with (5 females: Fig. 44), and the minimum size of unit that males are able to take over successfully (5 females: Table 17, and also see Table 23). This correlation is clearly not fortuitous, and it seems reasonable to conclude that the observed distribution of unit sizes in this population is a consequence of the fact that as units become larger than the critical size they begin to fragment socially and are therefore more easily taken over and broken up by males from the all-male groups. If loss of reproductive potential is the main functional reason why females in large units are more willing to desert, then we may expect the distribution of harem sizes to be set by the slope of this relationship. It seems reasonable to suppose that the upper limit on group size is set by the worst reproductive rate that females are prepared to tolerate, and we may doubt whether this is likely to be very great relative to the rate achieved by the highest-ranking females. Consequently, it seems unlikely that the slope could ever be very steep, for, if it were, females would be expected to desert sooner and the mean harem size would be smaller as a result.

One point of more general relevance may be noted. The structure of social relationships within gelada reproductive units is not constant: big groups are not small groups writ large, nor are small ones simply miniature versions of big groups. There is, rather, a demographically dependent shift in the structure of inter-individual relationships that can be seen to have profound consequences both for the individuals concerned and for the overall appearance of gelada society. With a few exceptions (e.g. Kummer 1968, Dunbar and Dunbar 1976), we have been inclined to assume that one group is much like any other group of the same species. Indeed, it has been an unwritten dictum in field studies of primates (in particular) that knowledge of what has been regarded as a fairly trivial inter-group variance in social structure should be sacrificed in the interests of a more finely detailed analysis of a single group. Platt and Denman (1975) comment on a similar tendency for ecologists to ignore the variance in their data and concentrate on the mean. While this tendency is often necessitated by logistic considerations, the present results should encourage us to broaden the scope of such studies and to be cognizant of the effects of changes in demographic structure on the behavior of animals (see also Altmann and Altmann 1979, Dunbar 1979a).

11 Rules and Decisions in Harem Acquisition

In this chapter, I summarize the options open to a male and examine the decisions he makes during the process of acquiring a harem. In the following chapter, I undertake a simulation analysis of harem acquisition strategies aimed at determining whether the alternative strategies that males pursue are equally profitable. Chapter 13 explores the tactics that males use to maximize the length of time they can retain control of their harems. In Chapter 14, I try to integrate the data on male harem acquisition strategies with the ecological and demographic data of Chapters 3 and 4 in order to explore the dynamics of the system as a whole. Finally, Chapter 15 returns to consider what happens to the male after he as been displaced as unit leader.

Strategies of Harem Acquisition

The male's initial problem in launching his reproductive career is to acquire control over a harem of females. He can do so in a number of ways, though only two are at all common.

The simplest strategy for acquiring a harem would obviously be to "kidnap" a loose female and form a small unit with her. Kummer (1968) describes something analogous to this in *Papio hamadryas*: in this species, young bachelor males kidnap infants and small juveniles and form "initial units" with them. Gelada males would evidently be happy to do this too, since a number of subadult and all-male group males were observed actively trying to interact with stray females, both adult and juvenile. In general, adult females rarely strayed. Juvenile females, on the other hand, strayed much more frequently, since they were herded less conscientiously by their males: they would frequently leave their units to interact with other juveniles and infants in the play groups that formed from time to time throughout the herd (Dunbar and Dunbar 1975, p. 98). Not only did these groups provide opportunities for young males to solicit passing juvenile females, but the males could expect to gain more or less unhindered access to such females away from the watchful eye of their harem-holders.

Interactions between subadult males and juvenile females were not uncommon in this context, although the males were apt to be threatened and the females herded whenever the harem-holder noticed what was going on (see also Mori 1979d, pp. 152–154). Nonetheless, some liaisons were more persistent. The subadult male of unit N12, for example, did at one point establish a grooming relationship (which included mounting) with the oestrous juvenile female of a neighboring unit, despite the attempts made by the females of N12 to drive her away. In another case, the subadult male of unit N19 was frequently observed grooming with the juvenile female of unit N14; they were also observed to copulate on several occasions when she came into oestrus. Neither relationship gave rise directly to a new unit, however.

While it is in theory possible that new units are formed in this way, it is probably rather rare in practice. This is likely to be so for two reasons. In the first place, harem males are certain to interfere in prospective relationships of this kind, thus making it difficult for a young male to build up any long-lasting bond with a female. In the second place, this bond would have to be strong enough to overcome the female's centripetal tendencies to remain with her female relatives. (The alternative that her relatives might desert with her once they perceived her intentions seems unlikely, given that the prospective new harem male is a subadult.) Nonetheless, such liaisons might be of fundamental importance in tactical terms for a male trying to enter units by more conventional means: his relationship with the female might well pave the way for his entry into her unit as a follower at a later date.

An alternative way of acquiring a harem would be to take over a unit that had lost its male. Presumably, harem males do, at least occasionally, die either by disease or accident on the cliff face while still incumbent. Since the harems do not split up if the male is removed, a bachelor male could move in and acquire the unit without too much trouble. At Gich, one harem-holder did disappear, leaving his unit intact, while another unit had its male experimentally removed (for details, see Mori 1979c, pp. 163–179). In the first case, another harem male took over the unit and tried to combine the two groups of females; his success, however, was limited, for the females strenuously resisted his attempts to combine them into a single unit. In the second case, a young adult male from an all-male group took over the unit during the harem male's enforced absence.

Although males may occasionally acquire units in this way, it is unlikely that a significant number of males can do so except in areas where male mortality is artificially high (e.g. Bole and other areas in the south of the gelada's range where males are shot for their capes:

see Dunbar 1977c). In undisturbed areas, the mortality rate for incumbent harem-holders was remarkably low. Of a total of 57 harem males observed over periods of 9–12 months at Sankaber and Gich, only one died while an incumbent. This is equivalent to a probability of death of only 0.025 per male per year, considerably lower than the overall death rate for adult males (0.133 per male per year on average). Since we would expect ailing harem males to be defeated and replaced by other males rather easily, instances of males dying incumbent are likely to be rare events.

A third strategy has already been alluded to: a male could challenge an incumbent harem-holder, defeat him, and take over the entire unit. This strategy is feasible because, as we have seen, large units become socially fragmented and unstable, with a concomitant decline in the loyalty of at least some of the females. At Sankaber, 7 takeovers were observed during 18 months of field study. Some of these are described in Dunbar and Dunbar (1975, pp. 106–113).

An interesting variant on this theme was observed at Bole. In this area, harem sizes were artificially large due to the periodic shooting of adult males. As a result, the few remaining males were apparently able to fuse maleless harems into unusually large units (Dunbar 1977c). Consequently, once the juvenile males had matured, they were able to break these large units up rather easily and detach small groups of females from them without having to defeat the incumbent harem-holder. One particular male was known to have lost females in this way to two different males while retaining control over the remnants of his unit on each occasion (see Dunbar 1977c, p. 373; also Dunbar and Dunbar 1975, p. 134). The readiness with which the Bole females apparently deserted their males is probably due to the fact that the harems were not only unusually large (see Table 1), but were also products of the fusion of several units. Mori's (1979c) observations on the attempts by one of the Gich males to fuse two harems suggest that groups of related females (remember that all the females in a harem are more closely related to each other than they are to the females of other units) do not like to be too closely associated with other groups of unrelated females. This seems reasonable in view of the way in which dominance ranks are determined, for the females of one unit are likely to be subordinated as a group to those of the other unit. The Bole situation seems to be a special case, and the possibility that males lose only part of their harems during takeover fights will not be considered further. It is worth noting, however, that Kummer's (1975) studies of artificially convened groups of unrelated gelada females produced rather similar

Table 20
All instances of the various stages of the follower cycle observed in the Sankaber main study bands in 1971–72 and 1974–75.

	Units Observed	
Stage	1971–72	1974–75
Follower enters unit	H63	N5, N16[a], N28
Established follower (immature social partners only)	H51	N5, N16[a], N28
Established follower with incipient unit	H39	N5, N16, N28
Unit Fission	H17, H63	N12

[a] Two instances observed in each case.

effects: when a second male was added to the compound, some of the lower ranking females deserted the first male for him.

There was also a fourth strategy that males adopted, namely, joining an existing reproductive unit as a "follower," then gradually building up relationships with some of the more peripheral females until a unit-within-a-unit (termed an *incipient unit*) had been established that could, in due course, split off to pursue an independent existence (see Dunbar and Dunbar 1975, pp. 103–106; also Mori 1979d). Although no male was ever observed right through the whole sequence, the various stages from joining, through owning an incipient unit, to unit fission were observed in a number of different cases during the two field studies. Table 20 lists all the instances in which each of the stages in this process were observed at Sankaber.

So far as I can see, there is only one other possibility that males could pursue as harem acquisition strategies, and this is for harems to be passively handed on from father to son as happens in hamadryas baboons (Sigg et al. 1982) and the gorilla (Harcourt and Stewart 1981). Although two of the followers at Sankaber did subsequently take their units over, in both cases they did so following a normal takeover fight. In neither case was there the slightest suggestion that the incumbent harem-holder was willingly handing over his females to his follower in the way that seems to happen in both the hamadryas and the gorilla. Pending further evidence to the contrary, both of these observed instances of takeover by a follower are assumed to be normal takeovers, and the possibility of nepotistic succession will not be considered further.

One final question remains to be considered, namely, what are the

Table 21

Frequencies with which males acquired harems by various strategies at Sankaber and Gich.

| Strategy | Frequency of occurrence[a] | | |
	Sankaber	Gich[b]	Total
Kidnapping stray females	—	—	—
Acquiring a maleless harem	—	1	1
Takeover	12	4	16
Follower-entry	15	4	19

[a] Includes both observed cases of harem acquisition and cases surmised from unit compositions (see text for details).

[b] Data given by Ohsawa (1979) and Mori (1979b,c,d).

relative frequencies of occurrence of the four strategies? We can answer this question in two ways: first, by considering all observed cases of harem acquisition and, secondly, by adding to these all surmised cases based on the compositions of individual units at the start of each study. We can use the compositions to determine the probable strategy of harem acquisition for all units that have at least two mature males, since any combination of a young male with an prime-aged male has to be a case of follower-entry, while any combination that includes an old male is fairly certainly the product of a takeover (see Dunbar and Dunbar 1975). This greatly increases the sample size, but does so at the expense of introducing errors of omission if the periods of time from which the two samples derive are significantly different. Thus, if males did not live very long after being taken over compared to the length of time that young males spent in reproductive units as followers, then units with young followers would be sampled over a longer time span than ones with old followers. (Remember that defeated haremholders remain in their former units as old followers: see Chapter 14, also Dunbar and Dunbar 1975.) Analyses given at the end of this chapter (see p. 144) and in Chapter 15 (p. 218) in fact suggest that old and young followers spend about the same length of time as members of their host units (2.1 and 2.3 years, respectively).

The data for the two analyses are given combined in Table 21, based on the units observed by us at Sankaber in 1971–72 and 1974–75 and by the Japanese at Gich in 1973–74 (based on Ohsawa [1979], using information given by Mori [1979b,c,d] to estimate the probable status of the second male). Although there was a slight preponderance of followers among the surmised cases, the distributions of observed and surmised cases are not significantly different ($\chi_1^2 = 0.569$, $p > 0.40$)

and the two sets of data may be pooled. The overwhelming majority of cases involved either takeovers or follower-entries, these being of approximately equal frequency (2×1 $\chi_1^2 = 0.257$, $p > 0.50$). The slight imbalance in favor of followers stems from the higher proportion of followers among the surmised cases: among the observed cases alone, there were 7 follower-entries to 8 takeovers.

Mechanics of Harem Entry

In this section, I describe in fairly general terms the sequence of events that takes place during a typical entry by a male into a reproductive unit. This summary is based on detailed observations at Sankaber of ten attempts by individual males to take over units and a further five males who entered units as followers. More detailed accounts of specific instances of entry can be found in Dunbar and Dunbar (1975, pp. 103–113). In the following section, I outline the essential features of this sequence of events as a set of decision rules and present evidence to support or test these rules where appropriate. In the final section, I pick up the follower's story from the point of entry into a unit and outline the process whereby he builds up his own harem of females.

Entry attempts begin when a young male from an all-male group associates closely with a reproductive unit, following it around and generally keeping fairly close to it without actually intruding into its social space. At this stage, it is not always possible to tell whether the male is attempting to join as a follower or is aiming at a takeover. A strange male obviously following a unit invariably invokes an aggressive reaction from the harem-holder and he will usually make a number of attempts to drive the male away. This will elicit one of three possible responses. Either the male does go away, or he responds submissively, cowering and snarling in response to all the harem-holder's attacks, or he will himself respond aggressively.

The first case is of academic interest only, for the entry attempt is thereby aborted. This response was observed only once (a young male attempting to enter N20, apparently as a follower).* Why a male should

---

* Even in this case, the male had successfully gained entry into the unit before he left. Moreover, before joining N20 he had briefly associated with another unit from a different band. A similar case occurred during the 1971–72 study: in this instance, a young male associated with unit H46 for a few days. One obvious interpretation of this behavior is that the males were assessing the profitability of the units. This is a possibility which I have not seriously considered in this account because of the paucity of data, but it does raise interesting questions.

behave in this way is unclear unless, of course, he was intending to try a takeover but thought better of it. The other two responses are more significant, for they provide the first clear indication of which direction the entry attempt will take. A submissive response leads to entry as a follower, whereas, by responding aggressively, the young male irrevocably commits himself to a serious fight over ownership of the unit.

If the intruder responds submissively, the harem-holder usually stops attacking and returns to his females. Although he will continue to threaten and attack the new male intermittently, the frequency of attack falls off rapidly and generally ceases altogether after a couple of days. During this period, the intruder will maintain a position on the periphery of the unit, conspicuously avoiding all attempts to interact with the adult members of the unit. His only efforts at interaction will be with the juveniles (usually males) who by the second day repeatedly go and sit near him with keen interest. This behavior often precipitates an attack by the harem-holder, but, as the frequency of these attacks declines, the intruder will begin to show interest in the juveniles and start to groom with them. Once these interactions occur, the intruder rapidly becomes an accepted member of the unit although, at least initially, his social interactions are restricted to immatures.

An aggressive response to the harem-holder's attacks forces the intruder into a fight, the outcome of which can only be one or other of the males retiring hurt, leaving hegemony over the unit to the other. Although, in effect, this is what happens, the actual process that brings about victory or defeat is much more complex and far from being a case of brute force and ignorance. From the outset, two completely different behavioral strategies are pursued by the males. The intruder's main problem is to get the females to interact with him; consequently, most of his effort is devoted to approaching individual females (who usually groom or feed during the mêlée) as closely as he can, staring pointedly at them and contact calling to them. The incumbent's task is clearly to prevent the young male from getting close enough to the females to interact: he spends his time trying to drive the intruder away whenever he comes too close to the unit. The intruder, if he is to have any hope of success, has to stand his ground as best he can (and, being younger, he is invariably smaller than the incumbent). Individual fights, though brief, are severe and injuries from canines are common: 8 out of 16 males involved in takeover fights suffered wounds, these being severe enough to cause the males to limp for some time afterwards in 4 of the cases (see Figs. 45–47). Fights are invariably broken off by the intruder, who may then circle round the unit to approach the females from another direction. As a result, fights can occur as frequently as

Figure 45   Male weaponry.

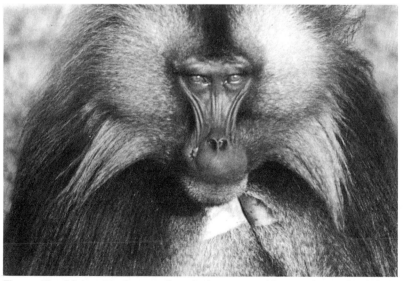

Figure 46   Male with deep canine slashes across his muzzle received during an unsuccessful attempt to take over unit N5 in May 1975.

Figure 47   Old male of unit H74 limping from a canine gash in his abdomen received during a takeover fight in November 1971. Note his generally disheveled and hunched appearance.

25 times per hour at peak intensities. Eventually (sometimes after several days of fighting) the outcome is decided either by the intruder giving up and returning to his all-male group or by a female interacting with him (Fig. 48).

Once the female has interacted with the intruder, others will soon follow; the incumbent then usually gives up and retires defeated to the periphery of the unit. This sounds rather prosaic, but is, in fact, a fairly precise description of what happens. Defeated harem-holders literally age overnight. Their chest patches fade from the brilliant scarlet of a harem male to the pale flesh-color typical of juveniles and old animals, their capes lose their luster and their gait loses its bounce. The changes are both dramatic and final. Kummer (personal communication) has noted similar rapid changes in the appearance of defeated unit leaders in *Papio hamadryas*.

What is crucial to the outcome is the behavior of the females. It is they who decide, by what amounts to a collective decision, whether to desert *en masse* to a new male or to retain their existing harem male. This decision is not made by any single female, but is "debated" and contested over a period of time by the females as a group. The intruding male will usually make passes at several of the unit's females before one of them starts to interact with him. Fighting (often as intense and severe as that between the males) then breaks out among the females.

Figure 48   Takeover fight for N28. (a) The former follower (*lower left*) solicits interaction with one of the females during his attempt to take the unit over; the harem male (*center right*) snarls and lip-flips in an attempt to prevent them from interacting, while the second follower male (*top left*) moves off to avoid the conflict. (b) The moment of success: the female hesitantly presents her rear to the former follower, while looking back at him; the second follower (*top left*) looks on while the harem male (*left*) backs off.

These fights seem to be genuine attempts by the females to interfere in the interactions between the intruder and individual females of the unit, and may last over several days before the issue is finally decided.

Neither fighting ability nor physical size is necessarily a good indicator as to which male, intruder or incumbent, will win. In at least some of the observed takeover fights, the incumbent seemed to be getting the better of the intruder yet lost the harem in the end. Conversely, the follower of the 1974–75 unit N5 repeatedly behaved submissively in response to attacks from the unit's two new males, and yet retained his female. What is important, nonetheless, is the intruder's ability to withstand the incumbent's repeated vigorous attempts to keep him away from the females. Loss of nerve at any time will result in instant failure to achieve that key goal of getting at least one female to interact with him.

Decision Rules of Harem Entry

The events described above can be set out more formally as a decision process with rules specifying each male's behavior. Some of these rules derive from observation, but others are essentially *a priori* and the evidence advanced in their support constitutes a *bona fide* test of the prediction.

The sequence of events is set out as a network of successive decision points in Figure 49. The pivotal decision-point lies in the intruder's response to the harem male's initial attacks. Submission commits a male to entering as a follower or to an immediate return to his all-male group.

Rule 1. An intruder should try to enter a large unit rather than a small one. We can deduce this rule from our knowledge of the structure of social relationships within gelada units. We know that, as harem size increases, the females interact less often with their males and are likely to be less loyal to him as a result. A small unit is likely to be too cohesive for the intruder to succeed in persuading a female to interact with him. Figure 50 plots the distribution of harem sizes available in the population during the two field studies at Sankaber, together with the distribution of harem sizes that males attempted to enter (irrespective of whether or not they eventually succeeded). A comparison of the distribution of entry attempts with expected values generated from the distribution of harem sizes in the population at large confirms that males did not attempt to enter units at random ($\chi^2_3 = 32.678, p \ll 0.001$).

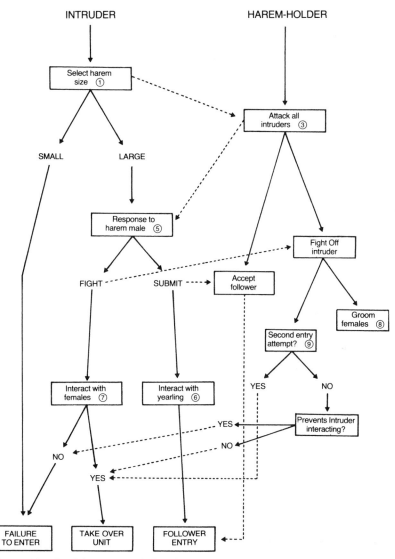

Figure 49   Decision network for the process of harem acquisition. Circled numbers in the decision boxes refer to the decision rules discussed in the text.

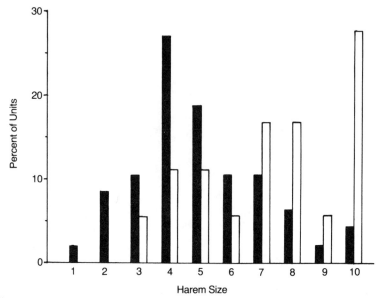

Figure 50   Distribution of the sizes of harems that males attempted to enter (whether or not they did so successfully: open bars, $n = 18$), compared to their relative availability in the sample population (filled bars, $n = 48$). Source of data: Main and Abyss bands in 1971–72 and Main band in 1974–75.

Inspection shows that males chose large units, and avoided small ones, more often than expected. Indeed, there is a strong negative correlation between harem size and the deviation of observed from expected values ($r_s = -0.945$, $n = 10$, $p < 0.001$), indicating that there is an increasing over-representation of units in the entries' sample as harem size increases. Contrary to what one might expect, males were not selecting for age of harem-holder, even though old males well past their prime might have been the easiest targets: the reason is that few males survive long as harem-holders and, if they do, they tend to have harems that are smaller than average which are therefore more cohesive (see Dunbar 1983b, Table I).

Rule 2. An intruder should leave his options open as long as possible. An intruder who commits himself at the outset to a takeover entry risks defeat if the harem male is exceptionally powerful and can keep him away from the females and/or the unit is unusually cohesive. Logically, the best strategy would be to keep a low profile and allow the harem male to open the bidding; that way, the intruder would get a brief

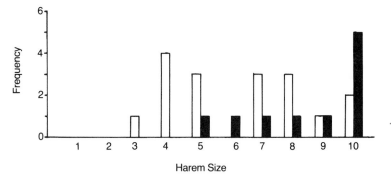

Figure 51 Distribution of harem sizes that males entered as followers (open bars, $n = 17$) or took over (filled bars, $n = 10$). The sample for followers includes data for 8 existing followers since the distribution for males who were already followers did not differ significantly from that for males who were observed making follower-entries (Kruskal-Wallis ANOVA: $H = 0.540$, df $= 1$, $p > 0.30$). Sample population as for Fig. 50.

glimpse of the harem male's capabilities before having to decide whether to try for a takeover, accept a follower-entry, or give up and try elsewhere. Unfortunately, it is rarely possible to observe an attempted entry from the outset: it is just not possible to monitor every unit throughout every day, and the first indication of an entry attempt is usually an outbreak of fighting somewhere in the herd. We could, perhaps, use the size of the units that followers and takeover males were observed trying to enter as an indirect test. The logic of the test rests on the fact that the ease with which a unit can be taken over increases exponentially with its size: we would therefore expect takeover strategists to select the larger units in order to maximize the likelihood of success, whereas followers should show no particular preference since entry can be effected just as easily to smaller units as to larger ones. However, this can at best be a one-way test: if the two distributions are similar, the hypothesis is supported, but if they are different we are really none the wiser since the males might have made rapid assessments of the situation before our arrival (see Rule 5 below). Figure 51 gives the frequencies of entry attempts into units of different size by followers and takeover strategists. Followers attempted to enter significantly smaller harems than did takeover males (medians of 5 and 9.5 females respectively; Kruskal-Wallis ANOVA, $H = 5.33$, df $= 1$, $p<0.05$). Nonetheless, the distributions overlap extensively, indicating that at least some followers attempt to enter units they could well have taken over. Strictly speaking, all we can conclude from this is that, at the latest, the males make a decision fairly soon after embarking on an entry attempt, but we cannot say whether this decision is made before or after approaching a target unit. I might be prepared to go a little further than this and suggest that, while some males probably do know which strategy they will pursue before they approach a unit, others almost certainly leave their options open. This is largely an intuitive conclusion, but it is based on the knowledge that at least two males who got into units as followers subsequently took those units over (one only a few months later). This implies that at least some followers are capable of pursuing a takeover strategy if the circumstances are propitious. It is likely that age/size is the key asymmetry prompting males to choose one strategy rather than another.

Rule 3. A harem-holder should always attack any male associating too closely with his unit. In prinicple, the harem male has no prior knowledge of an intruder's intentions. His dilemma at this juncture is that, if the intruder is ultimately aiming at a takeover, he should be vigorously discouraged, whereas, if the intruder's intention is to enter as a follower, he should be allowed to do so with minimal delay. (I will show in Chapter 13 that accepting a follower is one way of reducing the risk of takeover.) But, in these early stages, there is no unequivocal cue (from the intruder's behavior at least) that the harem male can use as a guide to the intruder's intentions. After all, there is nothing to stop an intruder from entering as a follower and then, once safely ensconced within the unit, taking it over from a position of greater strength. Indeed, at least two males did precisely this (units H74 and N28: for the former, see Dunbar and Dunbar 1975, p. 112). Clearly, the harem male's best strategy is to induce the intruder to declare his intentions, and the easiest way of doing this is to attack him, thereby forcing him to fight if he really is bent on a takeover. There was only one exception to this rule in 15 observed instances of attempted entry into a unit, and this was the case of the 5-year-old subadult male from H50 who entered unit H63 as a follower. The fact that the harem-holder did not attempt to discourage this male from associating with his unit may have been due to the intruder being unusually young and not, therefore, a significant threat. In all other cases, the incoming males were adult (6 years or older) and all where challenged by the harem male.

Rule 4. A harem male should suspect the intentions of an older male. In the following chapter it will be shown that males older than about 7 years can maximize their lifetime reproductive outputs only by taking over an entire harem in order to start their reproductive careers with as many females as possible. They can expect to do rather poorly if they try to enter a unit as a follower. Because of this, a harem male should be suspicious of any older male associating with his unit, since it is likely that he will want to take the unit over. Males continue to put on weight up to the age of 8–9 years, so that size gives a reliable guide to age. Conversely, a young male is likely to be aiming at follower-entry, and he should therefore be allowed to join the unit once his intentions have been tested (see Rule 3). Unfortunately, we have no satisfactory way of testing this prediction, since it requires data from the intruder's first moments of contact with the unit. It was, however, our general impression that older males who joined as followers were harassed much more by the harem male than were younger males. The young male who briefly joined N20 as a follower (see above) received only mild threats from the harem-holder, whereas the prime age males who joined N16 received much fiercer initial opposition.

Rule 5. To gain entry as a follower, an intruder must respond submissively to the harem male and avoid interacting with the females. A logical corollary of this is that any male who responds aggressively to the harem male's attacks *ipso facto* commits himself to a takeover fight. Although this rule was derived empirically, it can equally be derived as a logical consequence of the nature of harem takeovers and follower-entries. Of 11 successful entries to units that were observed during the two field studies at Sankaber, all 7 of the males who took over units responded aggressively to the harem male's attacks, whereas all 4 of the males who entered as followers responded submissively. Moreover, none of the males who entered as followers made any attempt to interact with the females, whereas all 7 of the takeover males did so. There were no known exceptions to this rule. Once a male has decided to attempt a follower-entry, his best strategy is to declare this as soon as possible in order to minimize the risk of precipitating a serious fight with the harem-holder. Any male who is initially undecided when he first approaches the unit (Rule 2) should make a rapid decision once he has been able to assess the situation at close quarters; Rule 5 thus places a time limit on Rule 2. This would explain why there was little unequivocal evidence to support Rule 2.

Rule 6. An intruder who opts for follower-entry will benefit by establishing a relationship with an infant at an early stage. This is an empirical rule. Of the 4 successful entries as followers, 3 of the males established a relationship with an infant or yearling at any early stage in the proceedings, while all 3 existing followers had such relationships

Figure 52   A yearling grooms a young follower. An early relationship with an infant facilitates social access to the reproductive females, and so paves the way to the establishment of an incipient unit.

when they were first observed. In all cases, these were positive associative relationships that involved grooming (Fig. 52). An infant is likely to be important to a follower for two reasons. First, this association provides access, in due course, to the females in the unit since, once a relationship has been established with an infant, it is a much easier task to establish one with the mother, and, through her, with the other females of the unit. Of two males who established relationships with infants, the first female with whom both subsequently interacted was the infant's mother. In two other cases, an established follower interacted with only one female and *her* yearling (H39 and N5: for the former, see Dunbar and Dunbar 1975, Fig. 25b). The second advantage of such a relationship is that the infant can be used as an agonistic buffer against the harem male (Fig. 53). Of the infants involved in encounters of this kind, 80.3% were male and 82.1% were in their second year of life (Dunbar 1984b, Tables 1 and 2). Use of infants in this way by followers is common throughout their sojourn in a unit.

Figure 53   A young follower carries a yearling on his back while under attack by his harem male. Note how tightly the infant clings to the follower; in many cases, the yearling initiates its own involvement in the encounter. A harem male invariably desists from his attack once an infant becomes involved.

Detailed descriptions of this behavior are given by Dunbar and Dunbar (1975, p. 53) and Dunbar (1984b). Carrying an infant is an extremely effective means of defusing the harem male's aggression. No attack on a follower was ever pressed home if it was carrying an infant. Infants often initiated their own involvement in an encounter by racing to the follower while he was being threatened by the harem male and leaping onto the follower's back. This suggests that the relationship between the follower and the infant must be a very close one. It seems plausible to suppose that the mother of such an infant must have a comparably strong bond with the follower if she is willing to allow her infant to be exposed to the risks inevitably involved in a fight. In functional terms, it probably means that the female will come to the follower's aid, if only to protect her infant from damage during any fighting (see Dunbar 1984b). Since this is likely to result in other females coming to *her* aid (see Chapter 7), the male faces the prospect of being attacked by some or all of the unit's females if he continues to press his attack on the follower.

Rule 7. An intruder attempting a takeover must interact with the unit's females. This is also an empirical rule: there is no *a priori* reason why we should predict that female choice is more important than male fighting ability in determining the outcome of a takeover contest. Nonetheless, we can perhaps see that it is a logical consequence of the particular nature of the female-female relationships in gelada combined with the effect of harem size on female loyalty. We have seen how important the female's decision is in terms of whether or not a new male can prise her away from her current male. Unless the intruder can persuade at least one female to interact with him, he cannot expect to win over the unit as a whole, and he will not be able to persuade any of the females to desert unless he puts considerable effort into soliciting them. Of the 10 males who responded aggressively to the harem-holder's initial attacks (thereby committing themselves to takeover attempts), all 10 actively solicited interaction with one or more of the unit's females. Seven succeeded in interacting with at least one female, and all of these subsequently took the unit over. The three males who failed to get close enough to interact with any of the females also failed to take the unit over.

Rule 8. A harem-holder under attack will try to strengthen his relationships with his females. Besides trying to prevent an intruder from interacting with his females, a harem-holder should try to prevent the females from deserting him by grooming with them during a takeover attempt. In fact, males under attack alternate between chasing off the intruder and frenetically grooming with their females (see Dunbar and Dunbar 1975, pp. 107–109). One point of interest is the attention that harem-holders under attack pay to black infants: they will pick them up and groom them to quite an unusual extent (see Dunbar 1984b). Such behavior in other species of baboons has been interpreted as a warning that the animal is prepared to invest considerably in the defense of his genetic investment (Busse and Hamilton 1981).

Rule 9. A harem-holder cannot survive repeated takeover attempts. Attempts by several males to take the same unit over are rather rare, partly, of course, because the majority of attempts are successful. However, of 10 observed takeover attempts, 3 failed. Two of these were against recently ensconced harem-holders, the third against an established male. Only in the latter case (N5) did another male attempt to take the unit over after the initial attempt had failed; in this case the harem-holder lost the unit. While the harem-holder had been able to keep the first male at a sufficient distance from the unit to make interaction with the females difficult, he was so exhausted from a day of continuous fighting that when, on the second day, another male began to challenge him, he was apparently unable to maintain a rate of counter-attack that was high enough to keep this male away from the unit: his rate of attack fell from 25.3 times per hour on the first day to 3.4 times per hour on the second. As a result, the second challenger's task of persuading the females to interact with him was that much easier. Harem males can do little to prevent further takeover attempts once they have successfully foiled one, since it is virtually impossible to cover up the noise and local disruption of a takeover fight. Most fights attract onlookers from the all-male groups. In the case of N5, the young male who made the second (successful) takeover attempt joined the proceedings as a spectator half way through the first day, but took no part in the fighting until the first intruder had retired back to his all-male group. Of 10 attempted takeovers, 6 had spectators from all-male groups who took a close, if non-participating, interest in the proceedings. One might be tempted to ask why bachelors do not form coalitions in order to take over large units that could then be split between them. There is no obvious reason why they should not do so, although it seems that they do not usually do it. The answer may lie in the fact that such large units are rather uncommon (see Fig. 5), while all males waiting to acquire harems are able to do so within a year (see Table 46). In other words, the rate of turnover is probably sufficiently high in this population to make it unnecessary for males to consider more forceful tactics.

Incipient Units and Harem Fission

Once a male has joined a unit as a follower, his only problem is to build up relationships with as many females as he can without infringing too seriously on the harem-holder's perceived hegemony over the unit. This does not appear to be an excessively difficult task, and most followers are able to form strong grooming relationships with at least one female fairly soon after joining a unit. In time, they may acquire more females: in two cases, the follower ultimately gained control over half the females in his unit. These relationships in due course extend to mating, although the harem male may sometimes try to prevent copulations by the follower. At this point, the follower and his female(s) constitute a distinct subunit (i.e. an *incipient unit*) within the group. Ultimately, the unit will undergo fission and the incipient unit will split off to adopt an independent existence of its own. The mean size of incipient units at the time of the fission of the parent unit was

4.0 reproductive females (range 3–6, $n = 3$). On average, exactly 50% of the reproductive females in the parent unit were under the follower's control at the time of fission.

A more extended discussion of harem fission can be found in Dunbar and Dunbar (1975, pp. 103–106).

We can estimate the length of time a follower spends in his host unit from the rate at which units with followers underwent fission. There were a total of 9 such units in the study bands at the start of the two field studies, and 3 of these underwent fission during the ensuing 9 months (giving an annual rate of 0.444 fissions per unit). The reciprocal of the annual rate is an estimate of the median length of "residence," and this turns out to be 2.25 years. Since this value is based on a relatively small sample (the 95% confidence limits for the probability of fission are $0.212 < p < 0.788$), too much emphasis should perhaps not be placed on it. Indeed, one male was known to have stayed on as a follower for 4 years without showing any signs of moving out. On balance, however, sojourns of about 2–3 years are probably typical.

12 An Economic Model of Male Reproductive Strategies

We have seen that a male has two main options in terms of harem acquisition and that these options commit him to lifetime strategies that are quite different. From a theoretical point of view, alternative solutions to the same problem are most likely to arise when the costs of pursuing the constraint-free strategy become so great that an alternative strategy is more profitable (Dunbar 1982a). The pay-offs of the different elements of the strategy-set may or may not be equilibrated, and it is this fact that often pinpoints the nature of the evolutionary explanation for the phenomenon. That males do pursue radically different reproductive strategies raises two questions. First, which, if either, is the more efficient in terms of reproductive output? And, second, why do some males choose one in preference to the other?

Males who pursue the two strategies could, in principle, generate markedly different numbers of offspring, but, because the frequencies of the males adopting the two strategies differ, the net contributions in terms of genes may be equal (Gadgil 1972, Maynard Smith 1974, Parker 1978). Consequently, the evolutionary benefits of strategies should, strictly speaking, be measured in terms of the relative numbers of genes contributed to the gene pool. However, where the strategies are essentially phenotypic and involve a degree of decision-making by the individual animal, then it is more likely that some intermediate criterion will be maximized (and hence equilibrated across strategies). Since, in contrast to females, male gelada are unable to influence the reproductive prospects of their offspring to any significant extent, it is likely that, if equilibration occurs at all, it will do so at the level of lifetime reproductive output. Equilibration at any one level is likely to result in *de facto* equilibration at all logically higher levels (where by "higher," I mean closer approximation to genetic fitness: see Dunbar 1983a). Consequently, lifetime reproductive output should provide a satisfactory estimate of genetic fitness.

In this chapter, I develop a model of male reproductive strategies in order to study their consequences in terms of lifetime reproductive output.

Assumptions of the Model

Lifetime reproductive output depends on two main variables, the birth rate per unit time and the length of time over which that rate can be maintained (see Dunbar 1982b). For a male gelada, this means in effect the number of females he has in his harem and the length of his tenure as harem-holder. In principle, therefore, all we need do to evaluate the relative efficiencies of the two strategies is to determine in each case (1) the mean size of the harem at the start of the breeding career, (2) the mean tenure as harem-holder and (3) the way in which harem size changes over time. If we know these, it is a relatively simple matter to determine the lifetime output for each strategy by simulation.

There are two other important assumptions, one logical and one strictly empirical. The first is that living in a social system based on one male reproductive groups obliges a male to gain control over such a group if he is to breed at all. The second is the observation that a male gets one, and only one, opportunity to own such a unit in his lifetime. These greatly simplify the problem, because only two relatively simple life-histories have to be considered. A male's "choice" of strategy is thus final, since, with one important exception to be discussed later, once the decision is made, he is committed to seeing it through. These two constraints, in particular, make it unlikely that equilibration of pay-off can occur at any level below lifetime reproductive output.

Strategy-Specific Costs and Benefits

In this section, I evaluate the baseline costs and benefits of the two strategies using data from the Sankaber population.

Costs

The strategy-dependent costs incurred by a follower seem to be trivial since he always avoids conflict. Entry, although initially opposed, is invariably successful.

In contrast, takeover strategists have to fight, often viciously, to have any hope of success; even then success is by no means guaranteed. It is important to remember here that sheer fighting ability *per se* is not the key to success in takeovers: it is ultimately the females who decide which male will win and which lose, and they do not appear to pay much attention to fighting prowess.

The incoming male's problem is to persuade the females to interact with him, while trying to circumvent the incumbent male's vigorous efforts to prevent him from doing so. Individual fights are normally of short duration, but are intense and frequent. These fights commonly result in wounds: 4 of 16 males involved in such fights received wounds severe enough to make them limp, while 4 others received more superficial injuries. Injuries may, on occasion, be so severe that a male is forced to retreat (1 of 10 males attempting entry). In other instances, males may be forced to retire to their all-male groups without serious injury simply because they have been unable to persuade the females to interact with them (2 more of the 10 cases). Thus, only 70% of the observed takeover attempts were successful. Failure to succeed may delay any further takeover attempt by that male, not only because of the need to recuperate from his wounds, but also due to the psychological disadvantage imposed by recent defeat: the male may be less confident and so less willing to risk or sustain injuries in a future contest. Ginsburg and Allee (1942) were able to show, in an elegant series of experiments with mice, that experience of defeat resulted in previously dominant males behaving nervously and responding submissively to males over whom they had recently been dominant.

Benefits

A male's reproductive output is, in principle, limited only by the number of females he can mate with. Males who attempt to take over units attack those that are significantly larger than the average for the population (Fig. 50). Consequently, these males can expect to do much better than followers who begin their breeding careers with one, or at most two, females.

The differential is not, however, quite as great as it might seem to be at first sight, for a male who takes over a unit will not necessarily gain *all* the females in it: some females may already "belong" to a follower in the unit, or he may lose some of the females to another male in a multiple takeover (see Chapter 13). The average number of breeding females obtained in 7 successful takeovers was 4.71, compared with 1.89 females "owned" by 9 followers (Fig. 54). The margin by which the instantaneous gain-rate of takeover strategists exceeds that of followers is reduced further by the fact that female reproductive

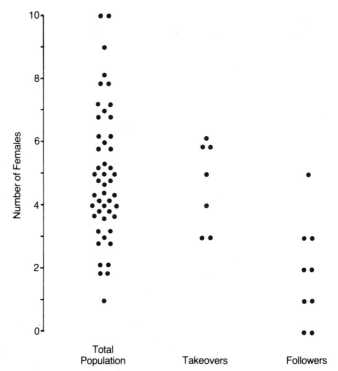

Figure 54　Number of breeding females acquired by takeover males and by followers, compared to the distribution of harem sizes in the population. Sample population as for Fig. 50.

Table 22
Strategy-specific baseline costs and benefits.

	Follower	Takeover
Mean number of breeding females at start	1.89	4.71
	(n = 10)	(n = 7)
Probability of wounding per attempt	0.00	0.50
	(n = 7)	(n = 16)
Probability of serious wounds/attempt	0.00	0.25
	(n = 7)	(n = 16)
Probability of success per attempt	1.00	0.70
	(n = 7)	(n = 10)
Mean age at start of tenure (yrs)	6.2	7.7
	(n = 16)	(n = 15)

rates decline as harem size increases as a consequence of the physiologically disruptive effects of social stress (Chapter 6). A follower cannot significantly increase his initial harem size at the start of his career, because doing so will tend to force him into conflict with the haremholder over the latter's hegemony over the unit. (He may gradually acquire more females by the time the unit undergoes fission, but these will have been acquired over a period of about 2 years.) Nonetheless, a takeover strategist can expect to have a rather higher instantaneous reproductive rate than a follower by virtue of the fact that he has more females.

Baseline Pay-offs

In summary, while the gains from a takeover are much greater than those from a follower strategy, the costs of the former are also greater. Nonetheless, since the risk of death from fighting is probably small, it is most unlikely that the costs of a takeover strategy outweigh its advantages. Even calculated crudely, a male can expect to gain only 1.89 females as a follower, but $4.71 \times 70\% = 3.30$ females from any given takeover attempt. The relative baseline gains and costs of the two strategies are summarized in Table 22.

Why, then, should males bother to pursue the follower strategy?

There are two quite distinct questions here. First, are there other factors that devalue the expected gains from a takeover such that, when calculated over a lifetime, they yield similar net gains? Secondly, are there factors that make individual males choose one strategy rather

Table 23

Relative frequency with which units were taken over in relation to their size.

Number of females	Number of units in sample[a]	Number of unit-months sampled	Number of takeovers
1	1	9	—
2	4	37	—
3	11	85	—
4	13	108	—
5	10	80	4
6	5	39	—
7	6	52	—
8	3	20	1
9	2	18	—
10	1	10	2

Source of data: Main and Abyss bands in 1971–72 and Main band in 1974–75.

[a] Units whose harem size changed during the study period are included at all observed sizes.

than the other? To answer these, we need to know a little more about the system constraints acting on a male.

Constraints on Male Behavior

GROUP SIZE AND FEMALE LOYALTY

We saw in Chapter 10 that, as the number of females in his unit increases, the male encounters a problem in time-sharing and is unable to interact with all his females with any regularity. This is partly a consequence of the limited time that a male has available for grooming with his females, and partly a consequence of the fact that the females themselves prefer to groom with close female relatives. The net effect is that a male is unable to interact with the excess females in his unit, and the group begins to become socially fragmented as the number of reproductive females increases above the "groomability" threshold.

From the intruder's point of view, this decreasing loyalty as unit size increases means that large units are easier to take over than small units. This is clearly borne out by the distribution of the sizes of units that males attempted to takeover (Fig. 51) and by the size-specific probabilities of successful takeover (Table 23; Fig. 55). The takeover rate is

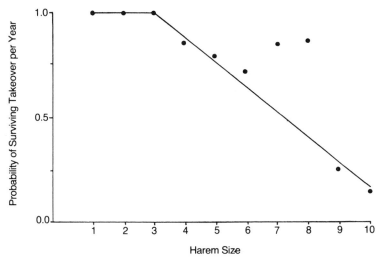

Figure 55   Probability that a harem male will not be taken over during the course of a year (3-point sliding mean of the data given in Table 23), plotted against harem size.

zero for units with 3 or fewer breeding females and increases as harem size increases.

These results have two important consequences for takeover strategists: (1) the bigger the unit, the easier it will be to take over and (2) having once taken it over, the new male will be at the same risk of being taken over himself. The probability of takeover is so high for large units, that a male has only a small chance of keeping it for any length of time. With the unit size of 10 breeding females, for example, his chances of surviving as harem-holder for 2 years are less than 2%.

MALE LIFE-HISTORY CONSTRAINTS

The second group of variables affecting a male's reproductive strategies are the consequences of his own life-history characteristics. Skeletal growth ceases at about the age of 6 years in male gelada (Dunbar 1980b), though they continue to put on weight until about the age of 9 years (see Kawai et al. 1983). Most harem-holders are older than 9 years, while the majority of males looking for units are younger. Prospective takeover males consequently tend to be at a size disadvantage vis-à-vis the males they are trying to displace. Although an incoming male does not have to defeat the incumbent male, he does have to be able to withstand considerable pressure from attacks by that male if he

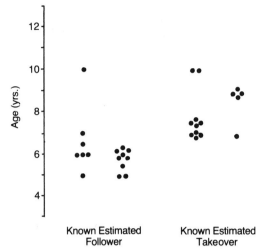

Figure 56   Ages at which males begin their breeding careers either as follow-ers or by taking over units. For both strategies, data derive from (a) a sample of males who were observed to enter a unit (known ages) and (b) a sample of males who were already established in a unit at the start of the study (estimated ages). In the latter case, the age of the male at entry was esti-mated from his current age in relation to the age(s) of the other male(s) in the unit, the degree of development of the male's social relationships, the ob-served periods during which defeated harem males and new followers re-mained in their units, and the interval between the two field studies. Source of data: Main and Abyss bands in 1971–72 and Main band in 1974–75.

is to be able to get close enough to the females to interact with them. One example of the fact that size affects fighting ability in gelada has been given above (p. 75), and it is a phenomenon that is widely re-ported from all taxonomic levels (e.g. *Papio* baboons, Packer 1979a; Bighorn sheep, Geist 1971; caprids, Schaller 1977, Dunbar and Dunbar 1981; frogs, Howard 1978; birds, Dunbar and Crook 1975; fish, Rub-enstein 1981; spiders, Riechert 1978). Figure 56 shows the distribution of the ages of males who were or became followers during our two field studies and that of males who attempted to take over units or had already done so. The estimated mean age of followers at the time of entry was 6.16 years (6.57 years for a sub-sample of 7 males who were actually observed to enter units as followers). The mean age for take-over entrants at the time of takeover was estimated to be 7.73 years (7.8 years for a sub-sample of 10 males who were observed attempting takeovers). The age distributions of the two classes differ significantly $\chi_4^2 = 20.869$, $p < 0.001$).

The second constraint is the male's own survivorship. His life expectancy sets an upper limit on the length of time for which he can hold a unit, and any delay in acquiring one after sexual maturity can only detract from his potential lifetime reproductive output. Thus, takeover strategists waste the better part of 2 full years of potential reproduction compared with followers. (We have no evidence to suggest that the survivorship function of followers is any different from that for takeover males, at least during the harem-holding period of their lives. I have therefore assumed that they are the same.) With an expectancy of life at physical maturity of only $e_6 = 6.59$ years (Dunbar 1980a), males have little leeway with which to gamble.

Strategy-Specific Lifetime Reproductive Success

To estimate the lifetime reproductive outputs of the two strategies, the expected number of offspring gained by each strategy was calculated iteratively over the successive years of the male's reproductive lifespan, given a specified female age distribution at the start. Female reproductive rates in each year were calculated on the basis of the age-specific fecundity rates given in Figure 13, these being adjusted to take into account the effect of harem size on female reproductive rates. (This was determined by using the regression equation for Table 6 to estimate the numbers of offspring in harems of different size and then setting a linear regression through the points so obtained.) The contribution to the strategy-specific lifetime reproductive output by males who survived to a given year of tenure was then determined by devaluing the number of offspring born each year by the probability that the male would survive as harem-holder into that year (i.e. by the conjoint probability of his not being taken over at his current harem size and of his surviving to the mid-point of that age class). At the end of each cycle, the female complement of the unit was adjusted to take account of female mortality. (Units do not normally change in size as a result of the migration of females, so that the number of breeding females in a unit was assumed to change only as a result of deaths and the maturation of female offspring: see Chapter 4). One further effect was taken into account in these calculations, namely, the likelihood that the unit would undergo fission. This is likely to have an important effect because it reduces the male's harem size, thereby reducing his reproductive rate while at the same time also reducing proportionately the likelihood of takeover (and thus prolonging his period of tenure). To incorporate this effect, I adjusted the age structure of the female complement of the harem in each cycle (i.e. year) by the likelihood of fission in the previous cycle ($p = 0.156$ per year for units with more than 3 reproductive females, independent of harem size: see p. 37), assuming that, when a unit did undergo fission, the follower took with him exactly half the females (see p. 74). This allows us to correct the age structure each year by assuming that the harem-holder has a probability of $p = 0.156$ of retaining only half as many females next year and $p = 1-0.156 = 0.844$ of retaining all of them. Hence, on average, he can expect to have $[0.5 \times 0.156] + [1 \times (1-0.156)] = 0.922$ of his females in the following year. This value is used as a scalar to adjust the female age structure of the harem at the beginning of each new cycle (or year), assuming that there are no age-dependent differences in the probabilities that individual females will desert with the follower or stay with the harem male. (This assumption is unlikely to be violated because females desert or stay as matrilineal groups, not as individuals.) Finally, the devalued reproductive outputs of males surviving to each year of tenure were summed across all years to yield the net strategy-specific expected lifetime reproductive output.

The functional relationships between the variables are summarized in Figure 57, and an outline of the computer program is given in Appendix B. The simulation used 6 sets of variable equations (given in Table 24): these were determined by setting least-squares regressions to the relevant data, as indicated in the final column of the table. Where the data clearly indicate a complex nonlinear relationship over the range of the x-variable, I have in general opted to fit a set of linear subequations rather than a single nonlinear one. The one exception is the equation for survival from takeover (equation 3): in this case, I have discounted the points corresponding to harem sizes of 6 and 7 females, and fitted a linear regression to the remaining data within the range $3 < n < 11$ (see Fig. 55). There is in fact a good reason for doing this. It is evident that, aside from these two points, the data fit a linear regression extraordinarily closely: this suggests that there is something anomalous about these two points. We will see in Chapter 13 that these points are indeed anomalous, at least from the point of view of our immediate concern, and that we are quite justified in omitting them. As a check on this, however, both linear and nonlinear equations will be fitted to the complete data set in the sensitivity analyses of the next section (see Table 25, equations 3.i and 3.ii).

The female and offspring complements of a unit at the start of each simulation were chosen to mirror the typical compositions of units of that size in the population (see Appendix B, Table A.2).

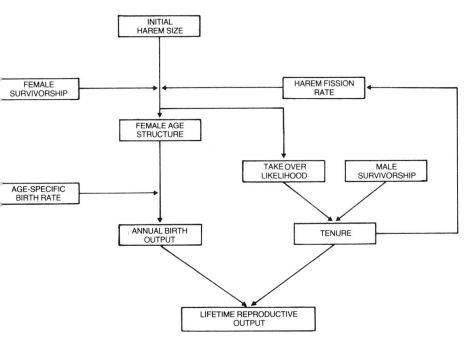

Figure 57 Flow diagram for the simulation of the lifetime reproductive outputs for the male's strategies of harem acquisition.

Expected lifetime outputs for males who started breeding at different ages with initial harems of various sizes are shown in Figure 58.

The results show quite clearly that, as starting age increases, output at a given harem size declines, as we would anticipate. More importantly, while a male can do absolutely better by effecting a takeover, to do so with advantage he must effect that takeover at the age of 6 years, an age at which he is probably physically incapable of doing so (see Fig. 56). The dotted lines in Figure 58 indicate the expected outputs for followers and takeover males with their respective observed mean number of females at the start of their breeding careers. The two strategies appear to be in almost perfect balance. Interpolating into the graphs using Bessel's formula for the difference calculus (see Pollard 1977) shows that a follower can expect to gain 4.724 offspring, while a takeover male gains 4.939. Although takeovers seem to give a male a slight advantage (the output for 6-year-old followers is 96% of the output for 8-year-old takeover males), with so many variables in the simulation estimated from data (each with its own estimation error) such a small difference can probably be discounted. That the outputs of the

Table 24

System equations used in the simulation of strategy-specific lifetime reproductive output for males.

Variable			Equations			r^2	Source
1. Age-specific fecundity	for	$x<4$	m_x	=	0		Fig. 13
		$3<x<6$	m_x	=	$0.2369x$		
		$5<x<9$	m_x	=	$0.1316x$		
		$8<x<13$	m_x	=	$0.1842x$		
		$x>12$	m_x	=	$0.0556x$		
2. Harem size fecundity effect			M_n	=	$0.5855 - 0.0238n$	0.999	Table 6
3. Size-specific probability of surviving takeover	for	$n<4$	S_n	=	1.0		Fig. 5E
		$n>3$	S_n	=	$1.3836 - 0.1231n$	0.985	
4. Male age-specific survivorship	for	$x = 6$	l_x	=	1.0		Fig. 11
		$6<x<9$	l_x	=	$1.1416 - 0.0229x$	0.895	
		$8<x<15$	l_x	=	$2.3224 - 0.1549x$	0.988	
		$14<x<18$	l_x	=	$0.4141 - 0.0234x$	0.937	
		$x>18$	l_x	=	0		
5. Female age-specific survival rate		$x<6$	q_x	=	$0.9585 + 0.0066x$	0.663	Fig. 11
		$5<x<17$	q_x	=	$1.1082 - 0.0186x$	0.997	
		$x>16$	q_x	=	$2.9810 - 0.1355x$		
6. Harem size adjustment due to fission			n_x	=	$0.922n$		Table 4

x = age class in 1-year intervals.
n = number of breeding females in unit.

two strategies turn out to be this close is quite astonishing considering all the computational steps.

Four important points should be noted. First, any delay in effecting a takeover reduces the male's expected output below that for all followers: by the age of 10 years it is already too late to do well as a takeover strategist. Conversely, the follower strategy is really viable only providing it is embarked on at an early age. By the time a male has reached the optimum age for a takeover, it is too late for him to do well as a follower. Thirdly, taking over too large a unit is counterproductive because of the high attrition rate on very large units. Finally, each curve has a shallow concave section within the range 3–6 females: this is due to the combined effect of takeovers and harem

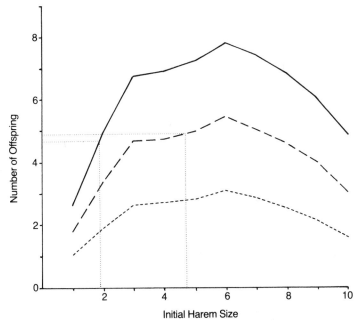

Figure 58 Results of the simulation of the lifetime reproductive outputs of males. The gross number of offspring produced by a male who started his breeding career at different ages is plotted against the size of his harem at the start of his breeding career. Three ages at start of breeding career are graphed: 6 years (solid line), 8 years (broken line), 10 years (pecked line). The mean lifetime outputs corresponding to the mean harem sizes for the follower (1.9 females) and takeover (4.7 females) strategies are indicated on the appropriate age graphs by the dotted lines.

fission that both become effective for the first time at a harem size of 4 females.

Sensitivity Analysis of the Model

While the basic model gives results that are consequences of the observed rates of behavior, we do not know how reliable these results are until we have checked the predictions of the model. We cannot, of course, check to see if the two strategies really do balance out in practice (there would have been little point in generating the model if we could have done so), and there are no other obvious predictions the model can make that are easily testable with the data available. An alternative approach to this problem is to determine how sensitive the model is to changes in parameter values. This will give us some idea as to how likely it is that the real lifetime reproductive outputs of the two strategies will differ from each other; or, rephrasing it the other way round, how different the parameter values have to be to yield significantly different lifetime outputs. More importantly, it will tell us which parameters are critical in setting up the conditions that make the two strategies equally profitable in the long run.

Altogether, I ran 12 sensitivity analyses aimed at examining the importance of each of the 7 variables involved in the model, in each case altering the value for just one variable. The variables were the female age-specific birth rate, the effect of harem size on birth rates, the harem fission rate, the takeover rate, the female age structure of the unit, and male and female age-specific survivorship. In addition, I considered the influence of initial harem size and the male's age at the start of tenure relative to these variables. In general, I looked at the effect of using an equation for the variable that was steeper or shallower than that used in the simulation. In the case of the female age-structure at the start of tenure, however, I considered the effect of having females that were older or younger than those used in the main simulation. The two other exceptions were the fission and harem size effects: in these cases, I simply looked to see what happened if the effect was removed altogether. Finally, I determined the consequences of a unit increase or decrease in the initial magnitude of the state variables (harem size and the male's age at entry). The equations used in the sensitivity analyses are given in Table 25. The final column of the table indicates the magnitude of the change in each case, expressed as the number of standard deviations by which the slope of the relationship has been changed. Most involve very substantial alterations.

The results of the analyses are given in Table 26 as the expected lifetime reproductive inputs for takeover males (aged 8 years with 4.71 females at the start of tenure) and followers (aged 6 years with 1.89 females at the start). The ratio of the two outputs (penultimate column) is given as a proportion of the output for takeover males in each case. The final column gives an estimate of the relative sensitivity of the model to changes in each variable: for this purpose, I use the mean percentage change in the ratio of outputs generated by a 1% change in the characteristic value (usually the mean) of the variable in question.

There are a number of points to note.

First, four of the variables (fission rate, age-specific birth rate, harem size effect on the birth rate, and female survival rate) have an insignificant effect on the behavior of the system as a whole (relative sensitiv-

Table 25

System equations used for sensitivity analyses of the model of male reproductive strategies.

Variable	Consequence	Equations		Magnitude of Change[a]
1. Fecundity	equalized across ages	$m_x = 0.1848x$	for $x>3$	−2.1
2. Harem size effect	effect removed	$M_n = n$	for all n	−1.3
3. Takeover rate	nonlinear[b]	$S_n = 1.0217 - 0.0306n$ $S_n = 6.9520 - 0.9011n$	for $3<n<9$ for $8<n<11$	−2.7
	$\bar{n}^c = 5.5$ vs. 7.2 females[d]	$S_n = 1.6000 - 0.2000n$	for $3<n<9$	
	$\bar{n} = 8.5$ vs. 7.2 females	$S_n = 1.3491 - 0.1025n$	for $3<n<11$	>5.0
4. Male survivorship	$e_6 = 9.4$ yrs vs. 13.2 yrs	$l_x = 2.0000 - 0.1667x$	for $5<x<13$	−4.2
	$e_6 = 15.5$ yrs vs. 13.2 yrs	$l_x = 1.3156 - 0.0526x$	for $6<x<26$	1.4
5. Female survival rate	$e_4 = 10.4$ yrs vs. 13.8 yrs	$q_x = 1.1280 - 0.0320x$	for $6<x<18$	−5.7
	$e_4 = 17.6$ yrs vs. 13.8 yrs	$q_x = 1.0925 - 0.0129x$ $q_x = 1.6781 - 0.0671x$	for $6<x<18$ $17<x<26$	>5.0
6. Harem fission rate	effect removed	$n_x = n$	for all x	−1.0
7. Female age structure	younger/older	set at start of simulation		2.6
8. Male's age at entry	younger/older	± 1 year from observed mean age		±0.9
9. Initial harem size	smaller/larger	± 1 female from observed size		±0.7

[a] Number of standard deviations by which the new equation differs from the old.
[b] Nonlinear regression fitted to the data given in Fig. 55 for $n>2$.
[c] Median size of unit taken over.
[d] Linear regression fitted to all the data given in Fig. 55 for $2<n<11$.

Table 26

Results of the sensitivity analyses of the model of male reproductive strategies, using the alternative equations specified in Table 25. The ratio of the two outputs and the relative sensitivity of the model to changes in each variable are given in the two final columns.

| Variable Altered | | Expected Lifetime Output | | Ratio (F/T) | Relative Sensitivity[c] |
		Takeover[a]	Follower[b]		
Basic model		4.939	4.724	0.956	
Birth rate		5.528	4.726	0.855	0.101
Harem size effect		5.247	4.784	0.912	0.046
Takeover rate	nonlinear	5.162	4.724	0.915	0.515
	n <obs[d]	5.259	4.719	0.897	
	n >obs	4.259	4.730	1.111	
Male survivorship	e_6<obs	1.886	2.577	1.367	0.879
	e_6>obs	9.065	8.266	0.912	
Female survival	e_4<obs	5.118	4.693	0.917	0.173
	e_4>obs	4.901	4.915	1.003	
Fission rate		5.083	4.723	0.929	0.028
Female age structure	$\bar{x}$<obs	5.624	4.702	0.891	0.324
	$\bar{x}$>obs	3.589	4.295	1.197	
Male's age at entry	<obs	6.030	5.028	0.884	0.615
	>obs	3.865	3.874	1.002	
Initial harem size	<obs	4.792	2.300	0.480	1.100
	>obs	5.343	6.683	1.251	

[a] Entry at 8 years with an initial harem size of 4.71 females.

[b] Entry at 6 years with an initial harem size of 1.89 females.

[c] Defined as the mean percentage change in the ratio of outputs (column 3) that results from a 1% change in the characteristic value of the variable in question.

[d] obs = observed value.

ity less than 0.2). In general, these variables have little effect on the reproductive expectations of followers, but increasingly depress the outputs of takeover males because these are older and have larger units (which are thereby more susceptible to fission). Note that, in general, the sensitivity indices are less than one: in other words, the influence of a change in the slope of a given variable on the dependent variable is damped. Only one independent variable (initial harem size) has a sensitivity greater than one.

Secondly, the remaining variables (male survivorship, takeover rate, initial harem size, female age structure, and male age at entry) all make significant contributions to the behavior of the system. Of these, male survivorship and initial harem size are the most important, with female age structure being the least significant. The takeover rate and female age structure act in similar ways to harem size and fission by affecting the output of older males with larger harems: they are of negligible importance to followers. Increasing the takeover rate (such that the median size of harems at takeover is reduced from 7.2 to 5.5 females) depresses a takeover strategist's output, while reducing it (to yield a median size at takeover of 8.5 females) has the opposite effect. Similarly, reducing the average age of females in the harem (thus benefiting from an improved age-specific birth rate and increased expectations of life for the females) increases the takeover male's output, while increasing the females' average age depresses it. Both variables, as with the three previous ones, exert their influence through harem size. The remaining factors affect both types of male equally, but do so unevenly. Altering any of these generally has a more significant impact on the expectations of older males, and the balance between the outputs of the two strategies is altered quite dramatically.

Finally, it will be noted that, in all of the latter cases, the consequences of changing the value of the relationship are not the same for the two strategies: reducing the slope gives a net gain to takeover males, but increasing it gives a net gain to followers. The significance of this is that the values estimated from the field data turn out to be just those that, in combination, result in an equilibration of pay-offs. The likelihood that this could be so by chance is clearly very small indeed. This point deserves emphasis since it provides what is perhaps the strongest evidence in support of the model's general validity.

These results suggest that there might well be frequency-dependent effects acting on the costs and benefits of the two strategies, such that if one of the variables is changed so as to benefit one strategy, the higher expectations of that strategy will encourage more males to pursue it; this would increase the competition for that strategy proportionately, and thus increase the cost. In fact, there are likely to be a number of equilibrium points, such that there will be a narrow zone of equilibration running across the n-dimensional hyperspace defined by the 9 variables in the system similar to the isoclines discussed by McFarland and Houston (1981). Some evidence of this is given in Table 26 in the case of increased male age at entry: here the combination of variables again yields a virtually equilibrated set of pay-offs. Thus, males may be able to offset costs on one variable by adjusting the values of some other variable(s) to regain an equilibration of pay-offs (cf. Dunbar 1983a).

Note that the effect of competition on cost may come through quite a different variable from that initially altered. Thus, increasing male survivorship would favor the takeover strategy, but the increased numbers of males pursuing that strategy would either result in an increased takeover rate (resulting in a reduction in mean tenure) or a reduction in the mean harem size at the start of tenure (as a result of multiple takeovers and, perhaps, higher fission rates). Thus, the immediate costs and benefits need not be along the same dimension, and hence will not be measured in the same currency; the ultimate costs and benefits, however, remain the same.

One final problem needs to be considered, namely, the accuracy of our age estimations in the field. Many of the underlying rate processes in the model are age-dependent, and if our ageing techniques are inaccurate, then the results of the simulation will be in error. Ageing primates in the field is notoriously difficult, and long-term studies have revealed that initial age estimates based on comparisons of physical size with well-fed captive animals often involve surprisingly large underestimates of a wild animal's true age (see Eisenberg et al. 1981, Sigg et al. 1982). There are two questions to consider: first, are our age classifications for the gelada accurate? And, second, if not, how seriously does this affect the results of the simulation?

There are three reasons for believing that our age classifications are reasonably accurate. In the first place, two groups of researchers working independently reached close agreement on the age ranges of the various size classes (see Kawai et al. 1983). Second, the age classes of the adults in particular were empirically determined from observations of long-term changes in the physical appearance of a small number known individuals at Sankaber (see Dunbar 1980a, Fig. 1). Finally, although these were checked against data for captive animals of known age (see Dunbar 1982c), it should be remembered that, unlike those populations of primates where large discrepancies in ages have been found, the Simen gelada do not suffer from serious food shortage (see Chapter 3). It seems unlikely that the age classifications used in this study are in error by any significant amount.

Even so, there are grounds for supposing that, if they are in error, the effect on the simulation of lifetime reproductive outputs is likely to be negligible. Of the 9 variables that enter into the model, only three will be seriously affected by errors in ageing criteria: these are the male's age at start of tenure, the male's age-specific survivorship, and the female age structure at the start of tenure. All the other variables are either not directly dependent on age (e.g. the harem fission rate) or are known to have a negligible effect on the results of the simulation (e.g. female survivorship). Of the three variables that are likely to be affected, errors in estimating ages will have the effect of increasing values on all three: males will have a longer expectation of life, but they will also embark on their reproductive careers at correspondingly greater ages with older females. The sensitivity analyses show that increasing the age estimates would have diametrically opposite effects on the variables: increasing survivorship gives the advantage to takeover males, but increasing the ages of the females and the male's age at entry gives the advantage to followers (see Table 26). The relative sensitivities of these two groups of variables are such as to just about cancel each other out. Simulation confirms that this is so. If the male's ages at entry are increased by 2 years to 8 and 10 years, respectively, male survival increased to give an expectation of life at the age of 6 years of 9.5 years (up from 7.2 years), the ages of the females increased by 2 years (with the age at first birth up from 4 to 6 years), and with the age-specific fecundity and survival schedules adjusted accordingly, the ratio of outputs is 0.906. Altering any one variable on its own results in output ratios that are very much more extreme (range 0.871 to 1.132).

In other words, providing our *relative* age classifications are correct, it does not seem to matter too much for the purposes of this simulation whether or not they are correct in terms of absolute chronological age, since the same errors are applied across the board.

Stability of Harem Size

Before concluding this chapter, one aside on the results of the simulations is worth making. We noted in Chapter 4 that fission probably serves to maintain the distribution of harem sizes at around a fairly stable mean. We can use the results of the simulation to show that this is in fact so. Figure 59 plots the mean harem size at the end of tenure (averaged across all male ages at entry) against the mean harem size at the start of tenure with and without fission. In the absence of fission, terminal harem size is approximately 17% larger than initial harem size (least-squares regression: slope $b = 1.171$, $r^2 = 0.972$, $t_8[b = 1] = 2.44$, $p = 0.041$). In contrast, fission results in a marked reduction in terminal harem size from these values: indeed, harems of 4 or more females (the minimum size for fission to occur) show a clear tendency to end up at a constant size of 4–6 females irrespective of their initial size.

This result is relevant to two empirical observations. First, it confirms the earlier conclusions that the observed overall mean harem size

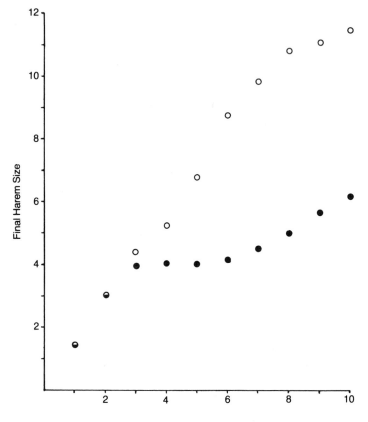

Initial Harem size

Figure 59 Influence of harem fission on mean harem size at the end of tenure, based on the results of the simulation of male reproductive strategies. Mean terminal harem size (averaged across all male ages at the start of tenure) is plotted against initial harem size in relation to whether harem fission occurs (filled circles) or does not (open circles).

of 4 females is the result of the fission process: the overall mean terminal harem size under the occurrence of fission for all initial harem sizes in Figure 59 is exactly 4.2 females. Second, it explains why old males were found to have harems smaller than those of males in their prime and of a similar size to those of young males who had only recently acquired their harems (see Dunbar and Dunbar 1975, Table 7; Dunbar 1983c, Table 1). We can now see that this is due to the fact that a male's harem size is increasingly likely to be reduced by a half as time passes.

13 Tactical Options
Open to Males

Although a male can expect to do as well by a takeover as by a follower entry, the values of the various parameters are by no means fixed. A male would gain a considerable advantage if he were able to manipulate any of them in his favor. In general, the initial values of most variables are extrinsically determined, being a consequence of various demographic processes over which a male has no control. Nonetheless, there are a variety of ways in which a male might alter the parameter values in his favor. The main tactical options he has available lie in either (1) adjusting his age at entry or the size of the harem he tries to take over, (2) capitalizing on opportunities that present themselves unexpectedly, or (3) trying to reduce the likelihood that he will himself be taken over once he has acquired his own harem.

Optimizing Strategy Choice

OPTIMIZING HAREM SIZE AT TAKEOVER

When a male takes over a unit, he is *ipso facto* at risk himself of being taken over, and the risks increase exponentially as the size of his unit continues to increase with time. A male would do best, other things being equal, to take over that size of unit which optimizes the ease with which large units can be taken over with respect to the survivorship advantages of small units in order to maximize the length of time for which he can hold on to the unit.

We can determine the optimum group size for any given length of tenure, t, by finding the group size that maximizes the expression:

$$p_i (1-p_i)^t,$$

where p_i is the probability of successfully taking over units of size i during a year and t is the length of tenure.

The average length of tenure for wild populations is not known, though at least one male was known to have held his unit for 4 years after he

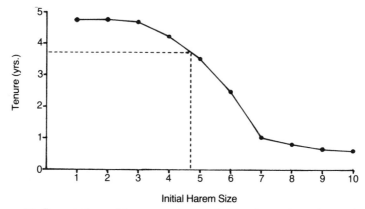

Figure 60 Expectation of tenure as harem-holder for males who took over units of various harem sizes at age 8 years, calculated from the conjoint probabilities of being able to take over a unit of a given size and of then being able to resist further takeover attempts. The mean harem size acquired by takeover strategists and the corresponding mean tenure are indicated by the broken lines.

took it over. However, tenure can be estimated from the takeover rates, and this can be done in two ways.

In the first place, we can determine how long it would take, on average, for all the units in the population to be taken over, given the observed takeover rate: this will give an estimate of the median tenure. During the 1971–72 study, 3 out of 14 large units (i.e. those with more than 4 reproductive females) were taken over, giving a rate of 0.214 per unit per year; taking the reciprocal of this, we obtain a median tenure of 4.67 years. During the 1974–75 study, 4 of 12 large units were taken over during a 9-month period, which is equivalent to an annual rate of 0.444 per unit; this gives a median tenure of 2.25 years. Taking an average of these two estimates gives a median tenure of 3.46 years (with 95% confidence limits of approximately 1.8 and 5.3 years).

A second estimate can be obtained from the reproductive output simulations by determining the median tenure for different initial harem sizes. These are graphed in Figure 60 for males who took units over at the age of 8 years. The expected tenure for a male with the observed mean number of females at takeover (i.e. 4.71 females: see Fig. 54) can then be determined by interpolation. This gives an expectation for tenure of 3.78 years, only slightly above the estimate obtained directly from the raw takeover rates.

Taking the average of the two estimates gives a tenure of about 3.6

Table 27

Probability of holding tenure at different harem sizes. The conjoint probability of being able to take over the unit and then resist takeover during a tenure of 3.6 years was determined from the size-specific probability of takeover (from Fig. 55).

Number of females	Probability of takeover/year	Probability of holding unit
1	0.000	0.000
2	0.000	0.000
3	0.000	0.000
4	0.108	0.072
5	0.231	0.090
6	0.354	0.073
7	0.477	0.047
8	0.600	0.022
9	0.723	0.007
10	0.846	0.001

years. The two estimates are not entirely independent, since they are both based on the same set of takeover data. However, they are calculated in completely different ways, using very different sets of additional information, so that they do provide some support for each other.

Inserting this estimate of tenure into the expression given above, we obtain a distribution for the conjoint probability of holding a harem with respect to unit size (Table 27). The maximum occurs at 5 females. The observed median harem size successfully taken over was, in fact, 5.0 females ($n = 7$, Table 23), while the mean number of females actually acquired by takeover was 4.7 (Table 22). If the conjoint probability distribution is used to generate an expected distribution for harem sizes at takeover, the observed and expected distributions are not significantly different $\chi^2_2 = 0.874$, $p = 0.646$).

Males may optimize the size of unit to take over with respect to maximizing their tenure, but does this necessarily yield the most offspring? Net output depends both on the number of females in the harem and the length of tenure, as well as the probability of being able to retain tenure for that length of time. Indeed, if we calculate the probabilities of retaining tenure at all harem sizes for a range of tenures we find that tenure retention is maximized at a harem size of 7 females held for just one year (Table 28). We can calculate the value of a given strategy to a male as the expected number of offspring, devalued by the conjoint probability of holding tenure—i.e. the number of females

Table 28
Probability of holding tenure for various periods of time given different harem sizes at the start of tenure for males who attempt to take over units (see text for details).

Initial Harem Size	Length of Tenure (yrs)					
	1	2	3	4	5	6
1	0	0	0	0	0	0
2	0	0	0	0	0	0
3	0	0	0	0	0	0
4	0.096	0.086	0.077	0.068	0.061	0.054
5	0.177	0.137	0.105	0.081	0.062	0.048
6	0.229	0.148	0.095	0.062	0.040	0.026
7	0.249	0.130	0.068	0.036	0.019	0.010
8	0.240	0.096	0.038	0.015	0.006	0.002
9	0.200	0.055	0.015	0.004	<0.001	<0.001
10	0.130	0.020	0.003	<0.001	<0.001	<0.001

times the birth rate (at the observed mean rate of 0.3396 births per female per year) times tenure duration times the strategy-specific probability of holding tenure (given in Table 28). While value is maximized at a harem size of 8 females held for a period of one year (Table 29), this is not necessarily the optimum strategy. A male cannot dictate how long he will retain control over his harem: the problem is thus equivalent to a single player game with no information. His optimum strategy in Game Theory terms is to choose whichever strategy minimizes his risk, and this will be that harem size which has the highest mean value when averaged across all tenure durations. This mean value is given in the final column of Table 29, and can be seen to reach its maximum at around 5–6 females. Six females is also the *maximin* solution (that which maximizes the minimum gain), so that this solution probably represents a saddle-point (i.e. is stable). Since the distribution of takeovers does not differ significantly from that predicted by the mean value $\chi^2_2 = 4.449, p = 0.108$), it seems as though males do make some attempt to optimize ease of takeover with respect to the probability of retaining the unit after takeover.

An alternative test is to note that, as the male's expectation of tenure declines, he should try to take over larger harems. Since the male's mean age at death is more or less fixed, the longer a male delays before taking over a unit, the shorter will be his tenure as harem-holder. A male aged 9–10 years can expect a tenure of only about 1 year, and

Table 29
Value of different combinations of tenure length and initial harem size to a takeover strategist. The tabled values are the weighted number of offspring produced by a given combination (i.e. total number of offspring produced assuming a constant birth rate of 0.3696 per female per year, devalued by the likelihood of holding tenure from Table 28).

Initial Harem Size	Length of Tenure (yrs)						Mean
	1	2	3	4	5	6	
1	0	0	0	0	0	0	0.000
2	0	0	0	0	0	0	0.000
3	0	0	0	0	0	0	0.000
4	0.142	0.254	0.342	0.402	0.451	0.479	0.345
5	0.327	0.506	0.582	0.599	0.573	0.532	0.519
6	0.508	0.656	0.632	0.550	0.444	0.346	0.523
7	0.644	0.673	0.528	0.373	0.246	0.155	0.437
8	0.710	0.568	0.337	0.177	0.071	0.035	0.316
9	0.665	0.366	0.150	0.053	0.020	0.007	0.210
10	0.480	0.148	0.033	0.077	0.001	<0.001	0.112

Table 30
Sizes of harems that males of different ages tried to take over.

	Male's Age (yrs)		
	6–7	7–8	9–10
Number of females in unit	5	5	10
	5	8	10
	5	10	
	10	10	
Mean number of females	6.3	8.3	10.0

Source of data: as for Table 23.

his best option is to go for the largest possible harem sizes (8–9 females); a male aged 6–7, on the other hand, can expect a tenure of about 4 years, and will therefore maximize his output by taking over harems with 5–6 females. Table 29 thus predicts a correlation between the male's age at the time of the entry-attempt and the size of the target harem. Although the sample size is rather small, this does in fact seem to be the case (Table 30: $r_s = 0.760$, $t_8 = 3.309$, $p < 0.02$).

Table 31
Observed distribution of the numbers of males attempting to enter large units
(i.e. those with more than 4 reproductive females), with an expected distribution
generated by a Poisson process with the same parameter (λ = 0.520 males
per unit) on the assumption that males attempt to enter units at random with
respect to each other.

Number of Units	Number of males attacking unit						Total
	0	1	2	3	4	5+	
Observed	19	3	1	1	—	1	25
Expected	14.9	7.7	2.0	0.3	(0.1)		

Source of data: Sankaber data: Main and Abyss bands in 1971–72 and Main band in
1974–75.

CAPITALIZING ON CHAOS

Takeovers are attended by a great deal of dissension among the females of the unit, partly because of attempts by some of the females
to prevent the desertion of the others (e.g. by threatening them when
they approach the incoming male to interact with him) and partly as a
result of the inevitable readjustment of female relationships vis-à-vis
the new male. Since it is difficult to conceal such a level of generalized
excitement, one might expect that other males would be attracted to
an unstable situation and so make attempts to take over the same unit
(see Chapter 11, Rule 9). That this is plausible is suggested by the fact
that in one case no fewer than 5 males attempted to take over one very
large unit within the space of two weeks, two of whom actually succeeded in dividing the unit's females between them.

To test the hypothesis, the observed distribution of the number of
males attempting to enter large units (i.e. those with at least four reproductive females) may be compared with an expected distribution
generated by a Poisson process with the same parameter, assuming that
males "decided" to enter units independently of each other's decisions.
Table 31 gives the observed and expected values for the combined data
from the main study bands of the two field studies at Sankaber. The
two distributions differ significantly (χ_1^2 = 4.210, $p < 0.05$). Multiple
entry attempts are much more common than expected, suggesting that
males do try to capitalize on the chaos and instability created by the
entry-attempts of other males.

Two points should be noted. First, two of the three multiple entry
attempts occurred in unusually large units. Second, this analysis refers
to all types of entries and not just to takeover attempts. I have included

attempts at follower-entry here since the entry attempts of other males may prompt a follower either to take over the unit (e.g. N28 in 1975) or to take the opportunity to acquire females earlier than he would otherwise have done (e.g. N16 in 1975).

BET-HEDGING

The male's problem reduces to one of whether he should become a follower now or wait in the hopes that he can take over a larger than average unit later on. One way out of this dilemma might be for the male to "hedge his bets" by entering a unit as a young follower with a view to taking it over at a later date if and when circumstances become more auspicious.

In fact, two males did do this: the follower of H74 took over the unit from the incumbent male after having been a follower for several months (see Dunbar and Dunbar 1975, p. 112), while the follower of N28 took his unit over when a second male joined it as a follower two months after his own entry. Although this represents 2/7 successful takeovers, it accounts for only 2/14 of all known followers (equivalent to an annual rate of transfer from follower to takeover status of about 0.190 per follower per year). At that rate, not only would most of the units have undergone fission (thereby rendering a successful takeover unlikely), but nearly half the followers would be dead before an opportunity to take their units over occurred. Although the 95% confidence limits on this point estimate are quite large ($0.103 < p < 0.748$), it is unlikely that the frequency of such takeovers could be very much higher than the observed rate. This is likely to be so because in most cases the followers of large units would often be beaten to it by an older male in a better position to take the unit over in the meantime (as in fact happened to the followers of unit H39 in 1972 and unit N5 in 1975). Thus, this tactic only seems to be viable on a strictly opportunistic basis, probably because only a small proportion of the units a male might join as a follower would be in a condition to be taken over about 2 years after he had entered it (i.e. at the time when he would be physically capable of taking it over).

DEMOGRAPHIC INFLUENCES ON STRATEGY CHOICE

It is inevitable that a male's choice of strategy will be influenced by the demographic structure of the population, if only in the trivial sense that he cannot take a unit over if there are no units in the population of a sufficient size to be taken over. The analyses so far, however, suggest

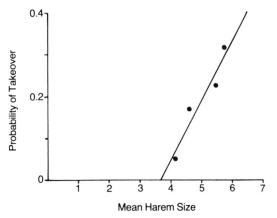

Figure 61 Probability of takeover per unit per year for 3 bands at Sankaber and one at Gich, plotted against the mean size of harems in the band. Regression line set by the method of least squares. Sample sizes are 25, 10, 17, and 6 reproductive units, respectively.

two more interesting hypotheses: (1) that the rate of takeovers will be related to the average size of units in the population (i.e. the more large units there are, the more males will be able to take units over) and (2) that the more large units there are, the greater the proportion of males who will attempt to enter units by a takeover rather than follower-entry. The second hypothesis, though obviously related to the first, does not necessarily follow from it, since successful entries do not necessarily follow from attempts at entry.

Figure 61 plots the probability *per annum* that a unit will be taken over against the mean number of reproductive females per unit for 4 bands. Two points may be noted. First, the relationship between the two variables is steeply linear ($r^2 = 0.918$; $t_2[b = 0] = 4.730$, $p = 0.021$ 1-tailed), indicating that the takeover rate rises rapidly as mean harem size increases. Second, extrapolating from the linear regression fitted to the data by the method of least squares, we find that the probability of a takeover becomes zero at a mean harem size of 3.66 females. Estimates of the actual size-specific takeover rates (Fig. 55) showed that this became zero at a harem size of 3–4 females. Conversely, the probability of takeover reaches unity at 10.80 females; the largest unit ever observed was one of 12 females ($n = 122$ units at Sankaber, 94 at Gich, and 10 at Bole).

Figure 62 plots the proportion of males attempting entry who did so by way of a takeover (as opposed to trying to become a follower) against mean harem size for the same 4 bands. Although the sample sizes are rather small, there is, again, a linear relationship between the two var-

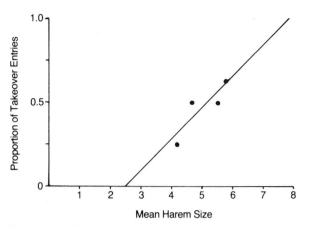

Figure 62 Proportion of males attempting to enter units per year who did so by way of takeover (as opposed to follower-entry) for 3 bands at Sankaber and one at Gich, plotted against the mean size of harems in the band. Regression line set by the method of least squares. Sample sizes are 4, 2, 2, and 11 males attempting entry, respectively.

iables ($r^2 = 0.763$; $t_2 = 2.540$, $p = 0.063$ 1-tailed), indicating that, as mean harem size increases, there is a corresponding increase in the proportion of males who attempt takeovers. Setting the regression equation equal to zero, we find that all entry attempts are by followers when the mean number of females per unit is less than 2.51; conversely, all entry attempts will be by takeover when the mean unit size exceeds 7.81 females.

It is worth commenting here that the demographic relationships were originally determined from the Sankaber data alone; they were then put to the test by using the Sankaber data to predict the values for Gich. The fits of the observed values for the Gich study area's Emetgogo band to those predicted by the data for the three Sankaber bands alone are satisfyingly close (Table 32). In both cases, the observed values are within the 95% confidence limits of the predicted values (calculated by the method recommended by Pollard 1977). In the knowledge of this result, however, I have proceeded to use all 4 data points in determining the regressions for Figures 61 and 62 on the grounds that, with such small samples, every extra data point gives a greatly improved fit to the underlying distribution.

MIGRATION TO ANOTHER BAND

One way a male could stack the demographic odds in his favor would be to transfer to a neighboring band that had a more disparate adult

Table 32

Comparisons of the observed rates of takeover and harem entry in the Gich Emetgogo band with those predicted by the regression equations for the Sankaber data.

	Probability of takeover/year	Proportion of males entering units by takeover
Emetgogo band (observed)	0.167	0.500
Predicted	0.119	0.345
t (df = 1)	−1.545	−2.675
p (2-tailed)	0.365	0.228

sex ratio (hence less competition for harems) or a larger mean harem size (hence higher probabilities of acquiring a larger than average harem by takeover). All-male groups are relatively mobile and often spend considerable amounts of time associating with other bands (see Chapter 2), and some males certainly acquire their harems in bands other than those with which they normally associate (presumably those into which they were born). Time spent associating with other bands would give a male the opportunity to assess his chances of acquiring a harem in a number of demographically very different bands, thereby permitting him to optimize his strategy choice. We cannot say whether or not males actually make decisions on this basis, though the opportunity to do so clearly exists. *Papio anubis* males, however, are known to transfer between groups in such a way that they gain a more advantageous oestrous-female/adult-male ratio (Packer 1979a).

We can use the migration rate data to estimate the proportion of males that acquire reproductive units in their natal bands by subtracting the known migration rate for males from the known rate of entry into reproductive units, since the difference has to be made up by males transferring between units within their natal bands. During two 9-month periods at Sankaber, 14 out of a total of 69 subadult and adult males that had not yet made their choice of reproductive strategy gained access to a reproductive unit, giving an annual rate of harem-entry of 0.236 per male (Table 46). During these periods, the inter-band migration rate for this class of males was 0.071 per male per year. Since, ultimately, all males acquire their own units (see Chapter 15), we deduce that 0.071/0.236 = 30.1% of the harem entries were by non-natal males and the rest (69.9%) were by natal males. Thus, most males appear to obtain their harems within their natal bands. We have no idea, however, what proximate factors might prompt an individual male to seek a harem in a band other than the one into which he was born. All-male groups do vary considerably in the extent to which they visit other bands (Dunbar and Dunbar 1975), and so to some extent a male's decision may simply reflect the experience he has gained through membership of a particular all-male group.

Counter-Strategies to Prolong Tenure

Given that a male has managed to acquire a reproductive unit, his problem becomes one of holding onto that unit for as long as possible. There is rather little he can do to increase the loyalty of his females by grooming with them more frequently (Chapter 10). Nonetheless, there are two other strategies that might serve to prolong the male's tenure as harem-holder. The first is for the male to reduce the size of his harem by shedding females; the second is to dissuade males from the all-male groups from choosing his unit as a suitable prospect for a takeover.

REDUCING EFFECTIVE UNIT SIZE

A male cannot shed females from his unit at will, however desirable such a stratagem might be, since the females are bound to their units by strong social bonds among themselves and the male has no direct control over their behavior (Chapters 5 and 7). It is an empirical observation that, with the exception of the very occasional juvenile, females do not leave their natal units to wander through the herd in search of new units (Chapter 4). Short of being taken over, the only way that a male can physically lose females is through their death (and we have absolutely no evidence to suggest that males might resort to *this* option).

There remains, however, one other possibility which, while not reducing *absolute* harem size, does reduce the male's *effective* harem size (i.e. the number of females who are bonded to *him*). He can permit another male to join the unit as a follower and allow him to take over the excess females. The effect will be to increase the average loyalty of the females in the unit, thus making it more difficult to take over.

Does the presence of a follower reduce the likelihood of takeover? The probability of takeover during a year (from Fig. 55) and the proportion of units of each size within the range 4–8 females that had

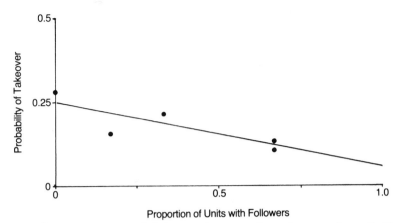

Figure 63 Effectiveness of followers in preventing takeover of a unit. The probability of takeover per year for harems of different size (from Fig. 55) is plotted against the proportion of units of that size which had followers, for units that were at risk of being taken over (i.e. had harems of size 4–8 females). Units with more than 8 females are not included in the analysis because the harem size is too large for the follower to be effective in the way hypothesized (see text).

followers are, in fact, negatively correlated, as predicted (Fig. 63; Spearman $r_s = -0.827$, $n = 5$, $p = 0.042$ 1-tailed). It is significant that the units with the largest proportion of followers were those with 7 and 8 females: this would explain the anomalous hump in the distribution of size-specific probabilities of takeover at this size (see Fig. 55). It may also explain the equally disconcerting hump in the mean number of offspring per female at the same harem sizes (see Dunbar and Dunbar 1977, Fig. 1): the females who groom with followers tend to be low-ranking, and these get a boost in their reproductive rates as a result of their association with the follower (see Chapter 8).

An alternative way to test the hypothesis is to compare the likelihoods of takeover for units that had followers and those that did not. Five out of 16 units that did not have followers were taken over (31.3%), whereas only 2 out of 9 units which had followers were (22.2%): the difference is statistically significant (Fisher exact test, $p = 0.008$).

It could be suggested that the advantage of a follower to the haremholder lies not so much in reducing his effective unit size as in the support which the follower can offer during takeover attempts. Allowing a follower access to a female might thus be a means of inducing him to join a coalition against outsiders through having a stake in the unit that he would lose if the unit were taken over. This quite plausible explanation is indistinguishable from the preceding explanation in terms of its final outcome (i.e. reducing the probability of takeover). However, the two differ considerably in the details of their behavioral mechanisms, for the second would require the active participation of the follower in takeover fights, whereas the first would not. In fact, in neither case where a unit with a follower was taken over (H39 and N5) did the follower play an active role in the proceedings; to the contrary, in both cases (or 3 cases, if the successive takeovers of unit N5 are counted as separate instances), the follower behaved submissively and did everything possible to stay out of the way. In both cases, the followers had incipient units of one female each; both retained their females after the takeover. Thus, this alternative explanation turns out to be irrelevant since the assumption on which it is based (that the follower will lose his stake in the harem following a takeover) is invalid. It seems that, if the strategy of taking on a follower is effective, it works because doing so reduces the effective harem size, thereby increasing the average loyalty of the females.

DISCOURAGING PROSPECTIVE CHALLENGERS

Could a male in some way indicate to prospective challengers in the all-male groups that he would be an unlikely candidate for an easy takeover, perhaps because his strength and prowess would allow him to keep contenders away from the females rather easily? There is one possibility, for harem owners and all-male groups engage in a ritualized threat-and-chase interaction ("yelping chases": see Crook 1966, Dunbar and Dunbar 1975) that would seem to be ideally suited to this purpose.

Typical interactions begin with a harem male leaving his females to run across to an all-male group in another part of the herd. While he stands threatening them, the males gather in a semicircle 1–2 m in front of him, threatening back (Fig. 64). After an agitated few minutes, the harem male turns and runs at a slow loping gallop back to his females, giving a high-pitched, two-phrase yelp that is unique both to the species and the context (see Dunbar and Dunbar 1975, p. 147), while the males of the all-male group follow in close pursuit, barking (Fig. 65). The pace is so slow that the all-male group males could easily catch the harem male: instead, they keep pace with him for distances of up to 100 m until he gets to within 15 m of his females, at which point they suddenly lose interest. These encounters elicit a great deal of interest from other males and are often contagious: harem males may follow

Figure 64 A harem male (*right*) challenges males from an all-male group.

one another in sequence, each in turn leaving his unit to initiate a chase as the preceding male returns to his females. The chases are invariably initiated by harem males, and only one harem male is ever involved at a time.

There are two possible mechanisms that might explain how this behavior discourages prospective challengers: either the absolute frequency with which a male engages in these chases is important, or, while engaging in a chase, the male is able to exhibit some sign that deters contenders (in which case, absolute frequency may be less important, providing all-male groups are "reminded" from time to time).

The first possibility is put to the test in Figure 66, which compares the rates at which all-male groups were challenged by males who were and those who were not subsequently taken over. (Samples for males who were taken over come from the months immediately prior to the takeover.) The two classes of males differ significantly (medians of 0.04 and 0.19 encounters per hr respectively; Mann-Whitney test, $p = 0.028$ 1-tailed): males challenged all-male groups significantly less often in the months prior to being taken over.

Although harem sizes differ between the two samples, there is no reason to suspect harem size as a confounding variable: given the doubtful loyalty of females in large harems, one would expect harem size to operate in the converse way in favor of more frequent challenging by

Figure 65 A yelping chase. Two young males from an all-male group (*center*) chase a harem male back to his unit. A female (*left*) looks unconcernedly on while feeding.

the males of large harems. In fact, there was a weak negative correlation between harem size and challenge rate ($r_s = -0.366$, $n = 9$, $p = 0.333$ 2-tailed). More plausibly, the age of the harem male might be important, in that males become "lazier" as they get older (cf. hamadryas baboon males: Kummer 1968): both of the males who were taken over were older males (>9 years old), whereas only 4 of the other 7 males were. If we consider older males only, the median rate of challenging for males who were not taken over was 0.24 per hr ($n = 4$), still clearly higher than the median rate of 0.04 encounters per hr obtained for males who were later taken over (Mann-Whitney test, $p = 0.067$ 1-tailed). Considered together, the median rate of challenging for older males who were not taken over was not significantly different from that for younger males who were not taken over (medians of 0.24 vs. 0.19, respectively; Mann-Whitney test, $p = 0.968$ 2-tailed).

It is worth noting that, of the older males who were not taken over, the two with the highest challenge rates both had large units (5 females each) at risk of being taken over, whereas the two older males with low challenge rates had small units below the risk threshold (4 females each). In contrast, the 3 younger males were much more consistent in their challenge rates. In other words, it is possible that, as males get

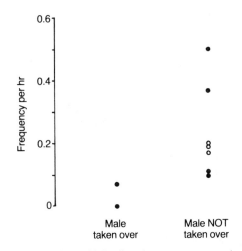

Figure 66 Frequencies with which all-male groups were challenged by harem males who were and who were not later taken over. Older males (solid circles) are distinguished from younger males (open circles). Based on timed samples (mean duration 1025 min) from 9 units studied in detail in 1974–75.

older, the rate at which they are prepared to challenge all-male groups does decline, but that males with large units at risk of being taken over (or males without followers) will step up the rate of challenging in order to delay the inevitable. Considerably more data would be required to evaluate these hypotheses further.

One final question remains: do all-male group males need to remember which harem males challenge them frequently?

In principle, it would not be difficult for the males to do so because they need to distinguish only among a small number of harem owners (25 in 1971–72, 17 in 1974–75), many of whom they will have known from infancy. Nonetheless, this would not be an option open to males immigrating in from other bands, and there seems to be no difference in the behavior of natal and immigrant males.

In fact, there seems to be no reason to invoke a sophisticated long-term memory, since males possess a means of displaying their effectiveness through the color of the areas of bare skin on the chest. In its "resting" state, this is usually a pale flesh color, but during aggression it turns bright pink (see Alvarez 1973). The chest patches of newly installed harem males turn bright crimson and remain so for some time, while the chests of defeated harem leaders lose their color almost overnight (see Dunbar and Dunbar 1975, p. 13). Chest coloration may thus reflect the frequency with which males are involved in agonistic encounters, not only with all-male groups but aslo with neighboring reproductive units, and may hence provide a rapidly assessed cue to the aggressiveness of individual harem males. Unfortunately, the data to test this hypothesis are not, at present, available. It does, however, suggest a reason why male gelada have evolved these areas of bare extensively venated skin on their chests. Indeed, it raises the possibility that the evolution of these areas of "sexual" skin could have come about as a result of selection on the male and not, as has always been tacitly assumed, through selection on the female.

Tactical Decision-Making

The tactical options open to males considerably complicate the simple two-strategy picture presented in Chapter 12. They imply that the males are assessing their options to a much greater extent, mainly by capitalizing on opportunities that arise. The simple two-choice decision that we studied in Chapter 12 thus turns out to have an overlying web of subtle tactical sub-decisions that give a male some freedom of movement in making his ultimate choice within the constraints laid down by the equilibration of average pay-offs for the two main strategies.

To evaluate every decision route for relative efficiency, as we did in Chapter 12 for the two main strategies, would be a hopelessly complex task for which very large samples of data would be required. In any case, it is not clear just how much we would learn from such analyses. The implications of these tactical options are that each individual male is able to adjust his behavior to optimize (within limits) his own life-history, and this means that each male would have to be assessed individually in the light of his particular circumstances.

The evidence for the existence of these tactical options does not in any way obviate the finding that the two main strategies are equilibrated in terms of reproductive output. The data from which the costs and benefits of those strategies were estimated incorporate the full variance in tactical options open to individual males. They show that, on average, a male can expect to gain about the same number of offspring either way, *providing* he pitches his age at entry and initial harem size correctly. What we have shown in this chapter is that if he fails to get his timing right, he can minimize the losses he will inevitably incur by a variety of decisions at the tactical level. One of these, for example, may be to adjust the size of unit he tries to take over to compensate for the reduced reproductive lifespan he will have if he delays his entry beyond the optimal age of entry for takeover males.

14 Dynamics of Strategy Choice

The analyses of Chapters 12 and 13 refer specifically to the Sankaber population. They have also taken a rather static approach to strategy choice in that they tend to assume that the biological context within which the male is embedded is effectively neutral, or at least constant over time. This, of course, is a gross oversimplification: environmental and demographic variables are in constant flux and influence each other in complex ways (see Chapter 4), and this is bound to have repercussions on the males' decisions (Dunbar 1979a, Altmann and Altmann 1979).

Since a male's strategic decisions are strongly influenced by the demographic structure of the population, he faces the invidious choice between entering a unit as a follower at age 6 years (with high chances of success) or delaying entry to try his luck at a takeover several years later (with a potentially high risk of being delayed by failure if the demographic structure of the population changes against him in the meantime). In the latter case, he may do better than the average follower if things are on his side, but he will do much worse if entry is delayed by any significant amount of time. In effect, the male is faced with the choice between a guaranteed average success by becoming a follower now (i.e. a low gain-rate with low variance) and the possibility of a potentially much greater success that is offset by the high risk of a low gain (i.e. a high gain-rate, but with high variance) if he delays to take over a unit in an unpredictable future. The risk a male is prepared to bear may become an important consideration (cf. Rubenstein 1982), although he may, as we saw in Chapter 13, be able to offset strategic misjudgments by tactical means to some extent.

In this chapter, I attempt to integrate our knowledge of the relationships between ecological, demographic, and behavioral variables with the results of the model of male reproductive strategies to show how, in the long term, the first two combine to influence the behavioral decisions of the males.

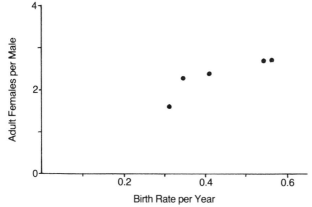

Figure 67 Adult sex ratio plotted against birth rate for 4 bands at Sankaber and one at Gich, based on data given by Ohsawa and Dunbar (1984, Table 1). Sample sizes are 65, 59, 118, 37, and 116 adults per band, respectively.

Interaction between Environment and Demography

Ohsawa and Dunbar (1984) identified a series of correlations that led from fundamental environmental variables (notably, ambient temperature, itself a simple function of altitude) through primary demographic variables (such as birth rate) to secondary demographic variables (in particular, the proportion of multimale units) that were a direct consequence of behavioral processes (in this case, rates of entry into reproductive units from all-male groups). Critical evaluation of the logically possible explanations for each correlation made it possible to draw firm conclusions about causal relationships at each step. Briefly, the sequence is as follows.

Altitude, as is well understood, determines ambient temperature, and this, combined with the effects of rainfall and high winds, creates a habitat-specific climate that becomes increasingly severe with altitude. This increasing climatic severity affects both birth rates and mortality rates. These two variables in turn influence the future adult sex ratio. The influence of birth rate on the adult sex ratio (Fig. 67) can be shown by simulation to be independent of the influence of climate acting via differential mortality rates (see Ohsawa and Dunbar 1984).

The ratio of females to males in the adult population determines the level of competition among males for access to females: the more females there are per male, the more likely it is that each male will be

Table 33
Demographic data for Simen bands.

Band		Number of units	Adult sex ratio[b]	Harem size ($\bar{x}$)	Male harem-holders (%)[c]	Males in AMGs[a] (%)	Multimale units (%)
Sankaber bands:							
Michibi	1971	20	3.0	3.7	80.0	8.0	15.0
Main	1971	25	2.7	3.4	81.3	6.3	23.1
Main	1974	17	2.4	4.9	48.6	20.0	35.3
Abyss	1971	6	2.7	4.5	60.0	10.0	33.3
Abyss	1974	11	2.3	3.7	61.1	5.6	36.4
High Hill	1971	12	3.7	4.6	80.0	6.7	16.7
High Hill	1974	12	2.8	4.7	60.0	25.0	16.7
E1	1971	3	1.8	3.0	60.0	0.0	66.7
E1	1974	5	2.5	5.4	45.5	9.1	60.0
E2	1971	2	2.4	6.0	40.0	20.0	100.0
E2	1974	5	4.0	4.8	83.3	0.0	20.0
Gich bands:							
Gich	1971	11	2.7	3.7	73.3	13.3	18.2
Khadadit	1971	14	2.9	3.5	82.4	11.8	3.4
Khadadit	1973	18	2.8	3.1	66.7	16.7	11.1
Emetgogo	1973	10	1.6	3.9	38.1	42.8	40.0
Emetgogo	1975	11	2.1	3.5	56.5	16.7	30.9
Frekyo	1973	3	2.0	3.3	60.0	0.0	33.3

[a] All-male groups.
[b] Ratio of adult females to adult males.
[c] Percentage of the adult males in the population as a whole.

able to acquire at least one female. Several consequences follow. The higher the proportion of males who become harem-holders, the fewer there will be in all-male groups waiting for the opportunity to acquire harems. This, in turn, will mean lower frequencies of takeover and follower-entry, and hence a smaller proportion of multimale units in the population. Table 33 gives demographic data for 17 Simen bands and provides evidence to support these conclusions.

We know that the rates of takeover and follower-entry are a function of mean harem size (Figs. 62 and 63). If these could be linked into the sequence outlined above, the result would be an extended functional chain that started with climatic conditions as the driving variable and ended with behavioral events and demographic structure as the outcome. Unfortunately, mean harem size cannot be reliably predicted from any demographic or environmental variable, but seems to be mainly the consequence of internal social processes (notably the rate at which units undergo fission). In fact, the causal network is rather complex here. This year's frequency of multimale units is a consequence of last year's rates of harem entry (which, in turn, depend on last year's mean harem size); at the same time, this year's frequency of multimale units determines how many units can undergo fission, thus determining next year's mean harem size. In practice, relating mean harem size directly to the percentage of multimale units gives the highest coefficient of determination of any combination of variables for a variety of linear and nonlinear regression models. Moreover, this procedure has the merit that, even if the causal sequence is somewhat garbled by linearizing a rather complex feedback loop, it does provide a predictive relationship that comes close to being statistically significant.

A Model of the Socio-Ecological System

The behavior of males at any given time can be seen to be the outcome of an extended series of interactions among a set of demographic and environmental variables acting at various times in the past. The relationships are summarized in Figure 68. For completeness, the flow chart also shows three loops that we have not yet discussed. Aside from these, however, it is clear that a linear causal chain can be defined. Since we can quantify the functional relationships involved, we can easily generate a simple linear "cascade" model of this component of the socio-ecological system.

In practice, of course, the functional relationships between the variables are rather more complex than suggested by Figure 68. We could build a literal demographic model using, say, Lesley matrices that replicated the actual dynamics of the system more closely. Aside from the fact that such a model would be extremely cumbersome, it has the serious disadvantage of losing generality since it would be necessary to model a particular population with a specific structure and composition. A simple linear model, in contrast, allows us to work in terms of generalized percentages.

Table 34 lists the regression equations obtained for each of the relationships shown in Figure 68. With a single exception, linear equations gave the best fit to the data. Although not all the regressions are statistically significant, the values of r^2 are generally high. They show

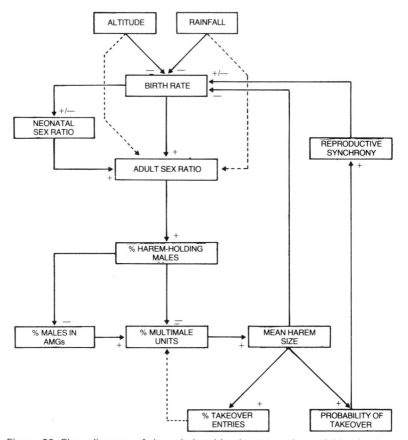

Figure 68 Flow diagram of the relationships between the variables in the gelada socio-ecological system. Positive and negative influences are indicated
by plus and minus signs adjacent to the arrowheads. The dotted lines indicate
causal relationships that are not formally included in the model.

that, with one exception, all the regression equations account for more
than 50% of the variance in the data to which they relate.

The exception is the relationship between the percentage of multimale units and mean harem size: this we already knew to be weak. We
could drop this relationship from the model, but doing so would intro-
duce another extrinsic (i.e. undetermined) variable that would have to
be initialized on each simulation. This would mean that the system's
output would have to be ascertained for all possible values of harem
size, as well as those for altitude and rainfall, thereby yielding an output in four dimensions that would be difficult to comprehend. Assum-

Table 34

System equations for the linear model of the gelada socio-ecological system.

Independent Variable	Dependent Variable	Equation	r^2	p[a]	Source
1. Rainfall (mm)	Birth rate (p/yr)	$y = 1.3541 - 0.000072x$	0.693	<0.10	Dunbar 1980a, Fig. 3
2. Altitude (km)	Birth rate (p/yr)	$y = 0.5682 - 0.0047\exp(x)$	0.788	<0.20	Ohsawa & Dunbar 1984, Table 1
3. Birth rate (p/yr)	Sex ratio[b]	$y = 1.0279 + 3.3197x$	0.996	<0.05	Fig. 68
4. Sex ratio[b]	Harem-holders (%)	$y = 17.521 + 17.542x$	0.513	<0.01	Table 33
5. Harem-holders (%)	Multimale units (%)	$y = 105.864 - 1.133x$	0.530	<0.01	Table 33
6. Multimale units (%)	Harem size ($\bar{x}$)	$y = 3.6249 + 0.0157x$	0.207	<0.10	Table 33
7. Harem size ($\bar{x}$)	Takeover (p/yr)	$y = -0.5132 + 0.1401x$	0.918	0.04	Fig. 61
8. Harem size ($\bar{x}$)	Takeover entry (p)	$y = -0.4736 + 0.1886x$	0.763	0.13	Fig. 62

Note: All equations were determined by fitting least-squares linear regressions to data from the specified populations.
[a] Significance level (2-tailed) of slope of regression, tested against the null hypothesis that $b = 0$.
[b] Ratio of adult females to adult males.

ing causal dependence between these two variables gives us the maximally deterministic relationship between the behavioral end-products and the underlying environmental variables: we need only remember that the variance introduced by stochastic effects in the harem fission process will make the end-products less predictable than they seem to be.

Note that, in constructing the model, I have short-circuited one other relationship, namely, the influence of climatic variables on the adult sex ratio via differential mortality rates. The main reason for doing this is that there are insufficient data from which to derive an equation relating the climatic variables and mortality rates. Simulation studies indicate that the interaction of differential mortality and birth rates predicts the observed adult sex ratios quite satisfactorily (Ohsawa and Dunbar 1984) and, since both are determined by the same driving variable (the climate), it seems reasonable to simplify matters slightly by proceeding only through the birth rate process. In any case, since the equations for this process are based on data that incorporate the effect via the mortality rates, the behavior of the system should not be seriously disrupted by doing this.

I have assumed a 6-year lag in the relationship between birth rate and its effect on adult sex ratio. Since climatic conditions actually determine conception and in utero survival rates rather than birth rates as such (see Dunbar 1980a), a 6-year lag is equivalent to the average of the maturation times for males and females (6 and 4 years, respectively) plus the lag time to the births following the onset of rainfall (roughly 9 months).

Linear Model with Two Environmental Variables

We consider first the effect of the two environmental factors (annual rainfall and altitude) on the demographic and behavioral variables, using the basic linear model outlined in Figure 68. The main elements of the computer program used in the simulation are given in Appendix C.

Tables 35 and 36 give the percentages of multimale units and the proportion of males who attempt to acquire harems by takeover at different altitudes under different rainfall regimes. The range on each of the two independent variables corresponds to that observed within the geographical range of the gelada. As the climate becomes more severe (higher altitude, higher rainfall), both the proportion of multimale units and the proportion of males acquiring their harems by takeover rise. These are consequences mainly of the fact that the birth rate declines as the climate becomes more severe, thus giving rise to more disparate adult sex ratios.

Table 35

Percentage of multimale units predicted by the linear model of the gelada socio-ecological system at various altitudes under different rainfall regimes.

Rainfall (mm)	Altitude (m)					
	2000	2500	3000	3500	4000	4500
1000	14.9	17.0	20.6	26.4	36.0	51.8
1100	20.7	22.6	25.7	30.9	39.4	53.4
1200	26.4	28.1	30.8	35.3	42.7	54.9
1300	32.2	33.6	35.9	39.8	46.1	56.5
1400	37.9	39.1	41.0	44.2	49.4	58.0
1500	43.7	44.6	46.1	48.7	52.8	59.6
1600	49.4	50.1	51.3	53.1	56.2	61.2

Table 36

Proportion of males attempting entry by takeover predicted by the linear model of the gelada ecological system at various altitudes under different rainfall regimes.

Rainfall (mm)	Altitude (m)					
	2000	2500	3000	3500	4000	4500
1000	0.385	0.392	0.403	0.420	0.450	0.498
1100	0.403	0.409	0.418	0.434	0.460	0.503
1200	0.421	0.426	0.434	0.448	0.470	0.508
1300	0.438	0.442	0.450	0.461	0.481	0.512
1400	0.456	0.459	0.465	0.475	0.491	0.517
1500	0.473	0.476	0.481	0.488	0.501	0.522
1600	0.491	0.493	0.496	0.502	0.511	0.527

Note that the percentage of multimale units varies from 15% to 61%, while the proportion of males attempting to acquire their harems by takeover varies from 0.38 to 0.52 over the same climatic range. The rate of takeover per harem also varies considerably (0.12 to 0.23 per year). These ranges are substantial and imply quite marked changes in both the demographic structure of the population and the behavior of the constituent animals. Note also that the variance in all the dependent variables declines with increasing altitude: in other words, birth rates are already so poor at high altitudes due to cold stress that additional stress imposed by an unusually wet year has relatively little effect.

How reliable is this model? Since we cannot check the model's predictions directly, we have only two options: (1) to carry out sensitivity analyses and (2) to check that the model is at least self-consistent (i.e. that it gives the correct predictions for the known data upon which it is based). For simplicity at this point, I shall opt for the second. In fact, it is a stronger test than might initially be supposed, since not all equations are based on data from each of the study areas.

Table 37 gives the observed and predicted values for the main dependent variables for the three study areas (with two sets of data for Sankaber and Gich corresponding to the two field studies at each site). One problem, however, is that we know the rainfall 6 years prior to the study year for only one of the five studies. As a first approximation, I have either used the mean rainfall or, where they are known, matched minima and maxima.

The fit to prediction for the percentage of multimale units is excellent in all studies except those at Gich, where the observed percentages seem to decline rather than continuing to increase as predicted by the model. Since this is consistent across two studies (and in fact is also true of our 1971 data for Gich), it seems unlikely that it is the consequence of any sort of random error. Clearly the present model fails at higher altitudes in this respect.

The fit to observed rates is much less good for the other two dependent variables. Both variables make passable predictions for two of the three Simen studies, but are wildly out for the Bole and Sankaber 1975 studies. The reason may well be that, while mean harem size is an excellent predictor of both takeover rates and the proportion of takeover entries, the percentage of multimale units is a relatively poor predictor of the mean harem size. This was a link we knew to be weak at the outset.

Nonetheless, that the predictions are of about the right size in two of the Simen cases is encouraging, considering the influence of internal social factors that may be quite independent of the external environmental variables. (Bole, being a disturbed population [see Dunbar 1977c], is best omitted from detailed consideration.) I will defer further consideration of this problem until the dynamic version of the model has been analyzed, since this will resolve many of the problems encountered with the linear model. For the moment, perhaps the most important point to note is that the model shows that very large inter-habitat variances in demographic and behavioral variables of the magnitudes actually observed can in principle arise as a result of the influence of relatively simple environmental parameters.

Table 37

Observed values for the three demographic variables in the three study areas, compared with values predicted by the linear model of the gelada socio-ecological system.

Study	Multimale Units (%)		Proportion of Takeover Entries		Probability of Takeover		Rainfall Basis
	Observed	Predicted	Observed	Predicted	Observed	Predicted	
Bole 1972	20.0	20.7	0.667	0.403	0.200	0.138	mean
Sankaber 1972	42.5	44.2	0.375	0.475	0.138	0.191	max
Sankaber 1975	33.7	35.3	0.636	0.448	0.314	0.171	min
Gich 1973	28.1	47.8	0.500	0.486	0.167	0.199	1968
Gich 1975	30.9	52.8	—	0.501	—	0.211	mean

DYNAMIC MODEL OF THE SANKABER POPULATION

With these provisos in mind, we now consider a more detailed dynamic model of the system in order to incorporate two important elements, namely, the feedback loops and the known fluctuations in annual rainfall. While medium-term cycles in rainfall of around 10 years in length are known to occur on a global scale (Pearson 1978) and, in particular, in this part of Africa (Wood and Lovett 1974), there is some evidence that shorter-term cycles of about 4-year intervals may be superimposed on these longer cycles (Dunbar 1980a; see Pearson 1978 for a more general discussion of short- and long-term weather cycles). If demographic and behavioral variables are determined by rainfall, then any cycles in the distribution of rainfall over time are bound to have significant and interesting consequences for the behavior of the animals. The possibility that climatic variation can have drastic affects on behavior has received little attention from biologists despite accumulating evidence (Winkless and Browning 1975, Pearson 1978). We shall consider two possibilities, namely, the effect of short-term cycles in the rainfall and the effect of randomly fluctuating rainfall. (The superimposition of a set of cycles of different periods often results in an apparently random sequence: see Platt and Denman 1975.) These two distribution patterns represent the extremes of the range of possible rainfall distributions. This model will be considered only with respect to Sandaber.

The equations used in the simulation (which follows the general pattern of Fig. 68) are listed in Tables 34 and 38, and the main elements of the computer program are given in Appendix D. We incorporate here the two feedback loops mentioned earlier, namely, the effects of harem size on birth rates (due to the suppression of reproductive activity in low-ranking females) and the tendency for females to come into oestrus prematurely following a takeover (thus giving rise to a local increase in the birth rate, followed by a subsequent drop the following year when those females would have given birth).

The equation for the first of these is given in Table 38, based on the relationship given in Table 6. This effect overlies the influence of climate on birth rate, as can be seen if the numbers of offspring in units of different sizes at Sankaber and Gich are plotted against the mean number of females per unit (Fig. 69). While the slopes for the three regression lines are very similar to each other, those for the two sets of Sankaber data are displaced significantly above that for Gich (Table 39). (Note that the test of the difference in slopes is further evidence, if proof were needed, for the hypothesis that dominance rank affects birth rate: see Chapter 6). Thus, birth rate can be seen to be a com-

Table 38

Additional equations used in the dynamic model of the gelada socio-ecological system.

Independent Variable	Dependent Variable	Equation	r^2	p^a	Source
1. Harem size ($\bar{x}$)	Birth rate (p/yr)	$y = 0.5855 - 0.0238x$	0.999	<0.001	Table 6
2. Takeover rate (p/yr)	Birth rate (p/yr)	(i) $z = 0.5(0.1455x - 0.3652)$ (ii) $y = z_i + M(1 - z_i - z_{i-1})$			
3. Birth rate (p/yr)	Neonatal sex ratio[b]	$y = 0.5028 \pm 0.1193x$	0.184	<0.30	Dunbar 1980a
4. Neonatal sex ratio[b]	Adult sex ratio[b]	$y = 10.0163 - 13.4989x$	0.918	<0.20	Ohsawa & Dunbar 1984

Notes:
(1) Equation 1 is the least squares linear regression fitted to the transformed values predicted by the regression of numbers of offspring on dominance rank (see Dunbar 1980b, Fig. 6); equations 2 and 3 are based on data for the Sankaber population; equation 4 on a simulation based on the demographic characteristics of the Sankaber population (see Ohsawa and Dunbar 1984).
(2) Equation 2 is calculated as the effect of the takeover rate (x) on the proportion of females brought prematurely into oestrus (z) in the current and the previous years, and the combined effects of these on the current year's gross birth rate (M) as determined by climatic factors (see Table 34).
[a] Significance level (2-tailed) of deviation of regression slope from zero.
[b] Ratio of females to males.

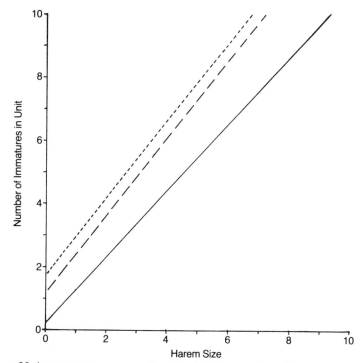

Figure 69 Least-squares regression fitted to the number of immature animals in 28 reproductive units of different size at Gich (solid line) (based on data given by Ohsawa [1979, Table 1.4] for the Emetgogo and Khadadit bands). Also shown are the regression lines fitted to the Sankaber data for 1971–72 (broken line) and 1974–75 (dotted line).

Table 39

Statistical analyses of the observed and predicted regression slopes of Figure 69.

Comparison	t	df	p (1-tailed)
Slope: Gich vs. Sankaber 1972	− 0.583	29	n.s.
Gich vs. Sankaber 1974	− 0.991	29	n.s.
Height: Gich vs. Sankaber 1972	− 1.390	29	<0.10
Gich vs. Sankaber 1974	− 1.793	29	<0.05

Table 40

Results of the simulation for the Sankaber population to show the influence of a cyclical pattern of rainfall on various demographic variables.

Year	Rain-fall (mm)	Birth rate (p/yr)	Sex ratio[a]	Harem-holding males (%)	Multi-male units (%)	Harem size (x̄)	Takeover rate (p/yr)	Takeover entries (p)
58								
59								
60	1200	0.490	2.19	64.2	42.5	4.29	0.187	0.469
61	1300	0.418	2.43	60.0	37.8	4.22	0.177	0.455
62	1400	0.347	2.66	56.0	33.2	4.15	0.166	0.441
63	1300	0.418	2.42	60.1	37.8	4.22	0.177	0.455
64								
65								

[a] Ratio of adult females to adult males.

pound function of climatic conditions and the distribution of harem sizes.

The second feedback loop cannot be specified by a simple equation, since it depends in part on the takeover rates in both the previous year and the current year. We know that about half the females come into oestrus prematurely following a takeover (see p. 96). We can, therefore, adjust the birth rate to compensate for the fact that half of the females in those units that were taken over have a birth rate of unity that year, and then do not contribute the following year. This is done by calculating a weighted mean birth rate for each year dependent on its takeover rate and its natural birth rate (as determined by the rainfall) and subtracting from this the proportion of females who would have given birth in that year had they not been induced to give birth the *previous* year by the fact that their units were taken over then.

The cyclical pattern of rainfall was defined by the sequence 1200–1300–1400–1300 mm of rainfall per year (roughly equivalent to the range observed at Sankaber). The effects of this sequence on the system yield the pattern of demographic and behavioral states illustrated in Table 40. The largest fluctuation occurs in the frequency of multi-male units, which ranges from 33–42%. The fluctuations in the proportion of males acquiring harems by takeover and the probability of takeover are much less marked.

For the random pattern of rainfall, the range was taken to be 1200–1400 mm as in the previous analysis, but the "actual" rainfall in any

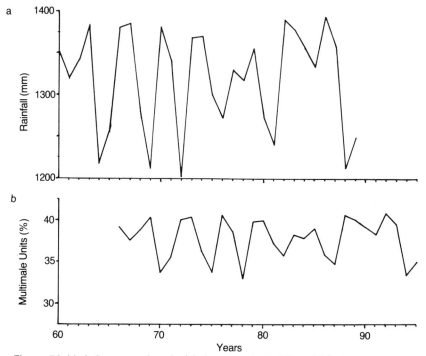

Figure 70 Variation over time in (a) the annual rainfall and (b) the percentage of multimale units predicted by the dynamic model with a random pattern of rainfall and no other stochastic effects.

given year was determined by a random number. The results are given in Figure 70 for the percentage of multimale units. The fluctuations in the dependent variable are clearly acyclic, though they are, of course, not greater than those predicted by a cyclical pattern of rainfall.

Up to now, we have assumed that the sex ratio at birth is exactly 50:50. While this is true in the long run, there are departures from this ratio, the magnitude of the departure in any given year being inversely related to the birth rate (Dunbar 1980a, p. 490). Since the neonatal sex ratio influences future adult sex ratios, small sample bias departures from an even neonatal sex ratio are bound to have important consequences for the demographic and behavioral variables. To incorporate the effect of random variations in the sex ratio at birth, I determined a relationship between the birth rate and the absolute deviance from a 50:50 neonatal sex ratio for the Sankaber data (see Dunbar 1980a, Fig. 2), and then determined a relationship between the absolute deviance in the neonatal sex ratio and the future adult sex ratio by simulation (using the model given by Ohsawa and Dunbar 1984). The equations for these two relationships are given in Table 38. The direction of the deviance from an even neonatal sex ratio in each "year" was determined by a random number.

Incorporating these two additional equations into the model results in quite dramatic variations in all three dependent variables. Note that the variance is of about the same magnitude for the cyclical and random rainfall patterns (Figs. 71 and 72). This increase in the variance is a consequence mainly of the random bias in the sex ratio at birth in years with low birth rates. This is the more remarkable given that in no one year was the deviance from an even neonatal sex ratio significantly different from 50:50 (see Dunbar 1980a). (Bear in mind, however, that, while a 60:40 distribution may not differ significantly from 50:50, it may well differ significantly from 40:60. Moreover, a succession of years biased towards one sex would yield significant departures from 50:50 in the adult cohort in due course.)

Rather more significant is the fact that the variance in each of the variables is now as great as that predicted by the linear model for all altitudes and all rainfall regimes, and is of the same magnitude as that actually observed at Sankaber (Table 41). Predictability from year to year declines sharply once these random factors are included in the model; this can be seen from the increase in Shannon's measure of information, H, which here measures the degree of predictability in the Markov chain matrix using lags of 2–10 years. Table 42 gives the values of H obtained for the percentage of multimale units. The fact that the cyclic and random rainfall regimes yield equally unpredictable outputs should be particularly noted, since it means that "chaos" results no matter what the actual pattern of rainfall is like.

Thus, our hesitation over the adequacy of the linear model due to its apparently poor predictability (Table 37) seems to have been unfounded. The lack of predictability was simply a consequence of the fact that, in making the predictions from the linear model, we did not take the dynamic properties of the system into consideration.

We have assumed that harem size is determined by the proportion of multimale units, when in act this is only partially true. Since harem size is also dependent on internal social factors that are quite independent of the system, a further degree of random bias is injected into the outcome. Thus, the system contains three quite different intrinsic sources of stochasticity: the climate, the neonatal sex ratio, and the frequency of harem fission. Between them, they make the demographic and behavioral outcomes unpredictable in the long term. We can make

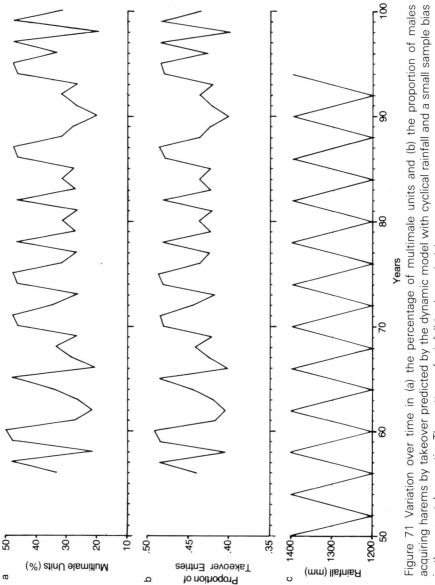

Figure 71 Variation over time in (a) the percentage of multimale units and (b) the proportion of males acquiring harems by takeover predicted by the dynamic model with cyclical rainfall and a small sample bias in the neonatal sex ratio. The pattern of rainfall is shown in (c).

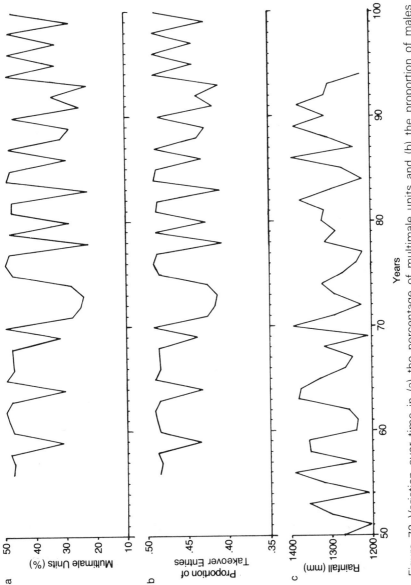
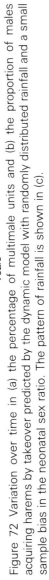

Figure 72 Variation over time in (a) the percentage of multimale units and (b) the proportion of males acquiring harems by takeover predicted by the dynamic model with randomly distributed rainfall and a small sample bias in the neonatal sex ratio. The pattern of rainfall is shown in (c).

Table 41

Ranges in the values of the demographic and behavioral variables predicted by the various models of the gelada socio-ecological system, compared to the ranges observed in the Sankaber population.

Model	Harem-holding males (%)	Multimale units (%)	Takeover rate (p/yr)	Takeover entries (p)
Observed at Sankaber	40.0–80.0	15.0–66.0[a]	0.138–0.314	0.375–0.636
Linear (all altitudes)	39.4–80.2	14.9–61.2	0.124–0.230	0.385–0.527
Dynamic: cyclical rainfall	56.0–64.2	33.2–42.5	0.166–0.187	0.441–0.469
random rainfall	57.2–64.2	33.2–41.1	0.166–0.184	0.442–0.465
cyclical rain with sex ratio bias[b]	50.7–75.5	20.6–50.2	0.137–0.205	0.403–0.493
random rain with sex ratio bias[b]	49.3–75.1	20.8–50.0	0.138–0.203	0.403–0.492

[a] Excludes the very small E2 band in 1971.
[b] Small sample bias on neonatal sex ratio.

Table 42

Predictability of frequencies of multi-male units at intervals of 2–10 years in the future using Shannon's index, H, for the various dynamic models of the gelada socio-ecological system (see text for details).

Model	Shannon's H	
	Mean	Range
Cyclical rainfall	1.193	1.039–1.385
Random rainfall	3.485	3.437–3.503
Cyclical rainfall with neonatal sex bias	2.947	2.811–3.160
Random rainfall with neonatal sex bias	3.732	3.624–3.801

useful predictions (i.e. ones that are likely to be right) only if we know the particular values for each of the key variables for any given spatio-temporally defined population. Otherwise, we can only predict long term *average* rates in which the sources of random bias due to birth sex ratio and the harem fission rate are averaged out; this would, of course, require comparably large samples based on many groups and/or many years against which to test these generalized predictions. The behavior of particular groups of animals, in contrast, would remain to all intents and purposes indeterminate. Somewhat analagous consequences have been pointed out by May (1976; May and Oster 1976) and Berryman

(1981) with respect to population dynamics. Simple deterministic causal relationships do not always yield simple, predictable results in complex biological systems, particularly when subject to small sample biases. That small sample biases can have important, and often dramatic, consequences for the behavior of animals has seldom been appreciated and is certainly never taken into account in studies of wild or captive populations (see Dunbar 1979a, Altmann and Altmann 1979).

Male Strategies in an Unpredictable Environment

Given that the demographic and behavioral variables fluctuate in an unpredictable way, how should a male behave? To assess the consequences for the strategic decisions of individual males, we need to take the model one step further and ask how the proportion of males acquiring units by takeover compares with the number of males waiting for the opportunity to acquire a unit.

The second of these two variables is equivalent to the percentage of males in all-male groups. This can be determined directly from the percentage of harem-holding males, since we know that these two variables are closely correlated (Ohsawa and Dunbar 1984). Setting a linear regression to the relevant data in Table 33 gives the following equation predicting the percentage of males in all-male groups from the percentage of harem-holders:

$$y = 38.720 - 0.420x,$$

with $r^2 = 0.342$ ($t_{15} = -2.699$, $p = 0.017$). The other variable (the proportion of males that will be able to take over units) can be determined by multiplying the percentage of harem-holding males by the probability of takeover.

The ratio of these two variables gives us the number of units available for takeover per male-waiting-for-the-opportunity-to-do-so. Figure 73 plots this ratio, derived from the random rainfall version of the dynamic model. It shows that in about one third of the years the number of acquirable units exceeds the number of males looking for harems, but fewer units are available than there are males the rest of the time. Thus, a male considering whether to enter a unit as a follower at the age of 6 years or to wait 2 more years to try for a takeover has a greater than even chance that the ratio of available units to prospective males will be unfavorable 2 years later. In fact, he will have a probability of 0.34 of a guaranteed takeover and a probability of 0.66 of uncertain success. The mean proportion of males waiting for units who

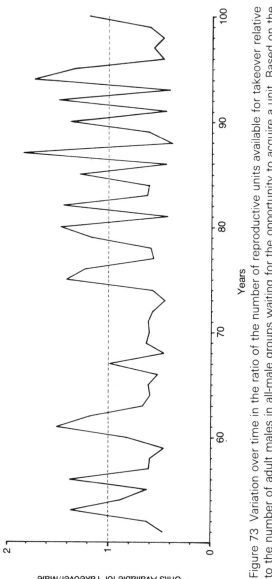

Figure 73 Variation over time in the ratio of the number of reproductive units available for takeover relative to the number of adult males in all-male groups waiting for the opportunity to acquire a unit. Based on the predictions of the dynamic model of the Sankaber socio-ecological system with random rainfall (see text for details).

will be able to acquire one in a "bad" year is 0.55; the remainder must either accept a reduced reproductive output as late-starting followers or delay for a further period to attempt a takeover at a later date. Those that opt for the second face the same uncertainties with a declining pay-off (see Fig. 59).

We can model the male's dilemma as a decision-tree, specifying at successive decision nodes the conditional probability of each outcome (Fig. 74). I have simplified the problem slightly by assuming that successive decisions can occur only at 2-year intervals, rather than in successive years (or, indeed, continuously in time). In effect, this is only to consider the decision process in terms of 2-year periods, and should make no difference to the outcome other than to reduce the grain of the analysis.

The decision sequence shown in Figure 74 runs like this. At the age of 6 years, a male chooses to become a follower with a probability p or to wait and attempt a takeover at age 8 years with probability $(1-p)$. At the age of 8 years, he has a probability of 0.34 of finding himself in a "good" year (and therefore successfully taking a unit over) and a probability of 0.66 of finding himself in a "bad" year (with a conditional probability of actually acquiring a unit by takeover of only 0.55). If he fails to acquire a unit, he must choose between becoming a follower (which he does with probability q) and delaying to try again at age 10 years (with probability $1-q$). At the age of 10 years, he faces the same decision, except that all males who fail to take a unit over in "bad" years must accept a follower-entry.

The pay-off for each decision can be obtained from the model of male reproductive strategies by interpolating into the graphs of Figure 59, assuming that the harem sizes at takeover are independent of age at entry.

We are left with two key unknowns, namely, the probabilities p and q of opting for follower-entry at ages 6 and 8 years, respectively. To determine these, we can cost out the expected pay-offs for each of the two strategies (takeover and follower-entry); then, assuming that the net pay-offs will be equal if the animals are behaving optimally, find the values for p and q that yield equilibrated pay-offs.

The net pay-off for followers is:

$$4.724p + 3.950(0.297q[1-p]) + 1.813(0.088[1-q][1-p])$$

and for takeover males is:

$$4.939(0.703[1-p]) + 2.813(0.209[1-q][1-p]).$$

Subtracting the second from the first, and setting the resultant equal to zero, we get:

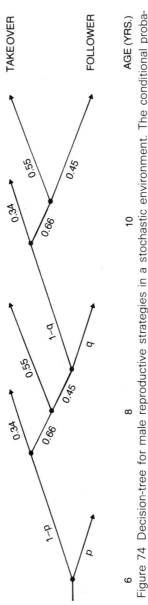

Figure 74 Decision-tree for male reproductive strategies in a stochastic environment. The conditional probability for each option at successive decision nodes is shown, assuming that males make their decisions at 6, 8, and 10 years of age.

Table 43

Relationship between the two unknown
decision probabilities of Figure 74.

q	p
0.0	0.452
0.1	0.442
0.2	0.431
0.3	0.420
0.4	0.408
0.5	0.396

$$4.724p + 1.173(q[1-p]) - 0.428([1-q][1-p]) -$$
$$3.472(1-p) = 0.$$

By rearranging, we can write an expression for q in terms of p:

$$q = (2.436 - 5.387p)(1-p)^{-2}.$$

Substituting for q in the preceding equation and multiplying out all the brackets, we obtain a value for p: it turns out to be $p = 0.452$. Substituting this back into the same equation then allows us to determine that $q = 0.003$. In fact, the value of p seems to be quite stable. If a range of values are assigned to q, p varies rather little (Table 43), which suggests that the underlying model is quite robust for modest errors in the estimation of its parameters.

These values can be used to determine the likelihood of each outcome in the decision-tree of Figure 74. Summing these gives a relative frequency of 0.501 for followers and 0.499 for takeovers. The probabilities of the two strategies vary only slightly in response to changes in the values of p and q. Taking the extreme values in Table 43, for example, yields frequencies for followers of 0.501 for [$p = 0.452$, $q = 0.0$] and 0.513 for [$p = 0.396$, $q = 0.5$].

In part, this robustness reflects the fact that we have assumed that the two strategies are equally profitable. We know this ought to be the case from the analyses in Chapter 12. The key question, however, is whether the decision model reflects the real world. This can be assessed by comparing the predicted distribution of decision outcomes with the frequencies observed in the natural population (given in Fig. 57). (Note that only the mean values of these distributions have featured in any of the models so far: since we are using the distributions themselves in this analysis, this is a *bona fide* test of the model.)

Table 44

Likelihood of each outcome in the decision-tree of Figure 74, together with the observed distribution (from Fig. 56) and the expected values generated by the likelihoods (given in parentheses).

Age at Decision (years)	Follower-Entry	Takeover
6	$p=0.452$	$p=0.000$
	14 (13.56)	0 (0.0)
8	$p=0.001$	$p=0.385$
	1 (0.03)	10 (11.55)
10	$p=0.048$	$p=0.114$
	1 (1.44)	4 (3.42)
Total	$p=0.501$	$p=0.499$
	16 (15.0)	14 (15.0)

Table 44 gives the probability of occurrence predicted by the model, together with the observed and expected frequencies, for each of the outcomes shown in Figure 74. The observed and expected distributions do not differ significantly $\chi_3^2 = 0.595$, $p = 0.898$). (Values for the two later ages of follower-entry are pooled due to the small expected values. The number of degrees of freedom in the censored 2×3 contingency table is 3: following Goodman [1968], one of the 6 cells is lost for the single parameter estimated from the data [the grand total], one for the cell lost by pooling, and one more for the cell that is *a priori* empty.) The robustness of the model can be checked again by re-evaluating Table 44 for extreme values of the two decision probabilities, p and q. In no case is the difference between the observed and expected values significant: in the worst case where $[p = 0.396, q = 0.5]$, $\chi_3^2 = 4.425$ ($p = 0.352$).

Nonetheless, it is possible that the apparent fit to prediction is a consequence of an increased likelihood of type II errors owing to the fact that we are here testing for similarity using relatively small samples. In fact, the results seem to be quite robust. We would have to increase the sample size more than 15 times to 450 entry-attempts in order to obtain a significant difference with the same proportional distributions of observed and expected values. Even with the observed sample size, a significant value for χ^2 would only be obtained if $[p<0.377, q>0.65]$. We can sidestep the problem by pooling cells to compare only the predicted frequencies of follower and takeover entries: at its worst, this yields $\chi_1^2 = 2.667$ ($p>0.10$), even for a 10-fold increase in the numerical size of the observed distribution. With the observed sample size, a significant value of χ^2 would be obtained if the proportion of followers was less than 37% or greater than 70%. In principle, it seems as though the model is robust in the general area of the predicted values (so that quite modest errors in parameter estimation and small sample biases in the data are unlikely to have serious consequences), but not so robust over its whole range that it could never be falsified by any data. The fact that the actual fit between observed and expected is so close (in fact, all but significantly *similar*) should give us considerable confidence in the results.

The excellent fit to prediction suggests that the decision model is basically correct. This in turn suggests that, at least in general terms, the males are behaving in a way that optimizes their reproductive decisions in the light of the uncertain future they face at each point in the sequence.

15 Two Final Problems About Males

Two questions have been left unanswered in the foregoing analyses. First, what becomes of the former harem-holders once they have been deposed? And secondly, is the set of alternative strategies evolutionarily stable? In this chapter, I try to answer these questions.

The Fate of Old Males

THE AFTERMATH OF TAKEOVER

In most species in which individual males hold harems that are taken from them by challengers, the defeated harem-holder either retires to an all-male group or moves elsewhere in the hopes of being able to take over another unit. Examples of both strategies are well known, not only in primates (langurs, Sugiyama 1965, Hrdy 1977; hamadryas baboons, Kummer 1968; patas, Hall 1965), but also in antelope (impala, Jarman 1979; hartebeest, Gosling 1974) and lions (Bertram 1975). The gelada are probably unique in that defeated harem-holders remain as followers in their former units.

The fact that defeated harem-holders stay on in their units is puzzling, since they thereby forgo any options they might otherwise have of taking over another unit (as many lekking antelope in effect do). This raises two problems: (1) why do the old males stay on? and (2) why do the new males tolerate them staying on?

Part of the answer to the first question seems to be that by the time a male has lost control of his harem, he is probably too old to be able to challenge another harem male with any success. Even if he did succeed, he would probably be unable to retain the harem for long in face of the competition for units that would inevitably result. On the other hand, an old male can contribute positively to his reproductive output by staying on in his former unit and helping to protect his offspring. Old males are extremely solicitous of the immature members of their units, especially the young infants; they are prepared to take on quite extraordinary odds in their defense. In these cases, the old males themselves frequently behave as though terrified; nonetheless, they pursue these challenges with a tenacity that is seldom seen in other contexts. These infants are, of course, their last (and final) reproductive contribution, and their continued protection at a vulnerable age is probably of considerable benefit in ensuring that these infants reach reproductive age. What is important, of course, is not so much the number of offspring produced, but the number of offspring who are themselves able to reproduce: evolution, to borrow a phrase, is interested in grandchildren, not in children.

There is, however, another consideration that may be of even greater importance. It has been argued (Angst and Thommen 1977, Hrdy 1979, Struhsaker 1977) that infanticide is a natural and inevitable phenomenon in all species where males hold harems of breeding females which they lose control over as a result of challenges by bachelor males (though Chapman and Hausfater [1979] have shown that males can gain by infanticide only under certain conditions of tenure length). The generally preferred functional explanation for infanticide is that the females who lose their infants come back into oestrus at once, and this saves the new male a considerable delay in starting his reproductive career. The male also may gain indirectly in terms of *relative* reproductive output by depressing the reproductive output of a rival. Females, in contrast, are generally considered to gain little and are in effect coerced against their interests by the males' superior physical power. If infanticide really is a universal phenomenon, then a male might be able to protect his last offspring from premature death by staying on and defending them against the new harem-holder. It is quite clear that old male gelada *do* defend their offspring against the new harem-holder and this might well deter any infanticidal inclinations that these males might have.

Nonetheless, it is odd that, while infanticide has been observed in a very wide range of primate and non-primate species (see Hrdy 1979), no species other than the gelada shows any evidence to suggest that old males attempt to protect their genetic investment. What circumstances make the gelada unique in this respect?

Costs and Benefits of Staying On

Two important features in the sequence of events following takeovers provide some suggestions as to what might be happening. The first is that the old males do provide some help to the new male in defending both the unit as a whole against encroachment by other units and the individual females against attack when the females encroach on the social space of neighboring units. They may also be of limited assistance to the harem-holder during takeover attempts by other males (as in the case of N5, where the two old followers repeatedly threatened the intruding males, albeit from a safe distance). In many cases, the younger reproductive females, as well as those that mature into the unit during the first few years after the takeover, are likely to be the old male's own offspring. Consequently, by affording them protection, he is contributing directly to his own reproductive output. The second point is that the females spontaneously come into oestrus following a takeover such that the mean length of post-partum amenorrhea is reduced from an average of 1.64 years to 1.14 years (see Dunbar 1980a, Fig. 6; see also Mori and Dunbar 1984). In other words, the females who would have come into premature oestrus as a result of the infanticide of their vulnerable offspring come into oestrus anyway (probably as a result of what Rowell [1978] has termed in patas monkeys the "Hoo Haa" effect). Thus, the main functional explanation for infanticide is removed.

It is interesting to note that each individual seems both to gain and to lose a little by this arrangement. The old male forgoes any possibility of acquiring another unit, but increases the likelihood that his existing offspring will reproduce successfully. The new male loses a potential gain in *relative* reproductive output, but gains a helper in the defense of the unit without incurring any serious cost in terms of delay in starting his own reproductive career. Finally, the females (or at least some of them) incur a cost in terms of a reduced probability of survival for their previous offspring as a result of coming into oestrus again prematurely, but gain by not losing one infant while at the same time stepping up their birth rate momentarily. In addition, the rapid return to oestrous cycling probably cements the bond between the females and the new male more quickly (see Kummer 1975), thereby reducing the social disruption caused by the takeover. This will, in turn, ultimately be reflected in improved reproductive performance all round.

In order to assess the relative value of the old male staying on to each member of this triad, we need to determine each member's expected gains from this event, together with what is technically known in Game Theory as its "regret" (that is, what it forgoes by not pursuing the alternative strategy). Substracting the second from the first gives each individual's net gain (positive or negative) due to the old male staying on and the new male refraining from infanticide. Table 45 costs out these various gains.

The least certain case is for the female. (Note that we only need to consider the case of a female at risk of losing an offspring by infanticide: a female who does not have such an offpsring will have a much

Table 45

Estimated gain and regret due to the new harem-holder refraining from infanticide and the old male staying on in the unit (see text for details).

	Gain Source	Offspring	Regret Source	Offspring
Female	premature cycling	= 0.233	premature cycling	= 0.233
	saved from infanticide	= 1.000	loss due to infanticide	= −1.000
	risk to old infant	= −0.000	risk to new infant	= −0.004
	risk to new infant	= −0.004	risk to self	= −?
	risk to self	= −?		
	TOTAL	= 1.229−?	TOTAL	= −0.771−?
	NET GAIN	= 2.000		
Old male	saved from infanticide	= 3.250	takeover of new unit	= 1.555
	aid to daughters	= ?	infanticide on first unit	= −3.250
	TOTAL	= 3.250 + ?	TOTAL	= −1.695
	NET GAIN	= 4.945 + ?		
New male	female premature cycles	= 2.635	infanticide to rival	= 3.250
	support from old male	= ?	female premature cycles	= 2.635
	TOTAL	= 2.635 + ?	TOTAL	= 5.885
	NET GAIN	= −3.250 + ?		

less specific interest in the issue.) Her gain is the extra number of off-spring she can fit into her reproductive life by coming into oestrus 6 months prematurely. With a birth rate of 0.467 per year (equivalent to the mean inter-birth interval of 2.143 years), she will gain an extra $0.5 \times 0.467 = 0.233$ offspring. Estimating the increased mortality among infants due to early weaning is not easy since none of the infants con-cerned were followed through to reproductive maturity. However, 21 infants and yearlings aged 0–18 months at the time of takeover were followed for 2–12 months after the takeover and no deaths were re-corded. Since this is likely to be the infant's most vulnerable period, and the natural mortality rate over the period 6–24 months is of the order of 0.002 per animal per month, a mortality rate of zero in the takeover sample at least suggests that this cost is not significant. By comparison, to outweigh the gain of 0.23 infants over a lifetime due a single instance of early weaning, the mortality rate would have to be 0.010 per infant per month, an increase of nearly 500%. Clearly con-siderable marginal increases in the mortality of early-weaned offspring are possible without offsetting the gain accrued from a premature re-turn to breeding condition. However, Mori and Dunbar (1984) found that females with very young infants (less than 6 months old) rarely came into oestrus prematurely, suggesting that the risks to such young infants may be too great to warrant an early return to reproductive condition. (Recall that 6 months is the point at which the female's energy investment in her offspring starts to decline as the infant begins to feed for itself: see Fig. 32).

It is equally difficult to estimate the costs of premature conception in terms of the reduced survivorship of the mother. An alternative approach is to determine how great the mortality rate for females would have to be to offset their net gain from premature cycling. The females' gain is the extra number of offspring born by cycling prematurely (0.23) plus the offspring saved from infanticide, less the number of new and old infants that die as a result of their mother's premature cycling (0.004 in all). To offset the net gain of 1.23 offspring during the remainder of her reproductive life, the average female (i.e. one that was exactly halfway through her expected reproductive lifespan of 9.8 years) would have to die 3.346 years earlier than anticipated: this is the time re-quired, on average, to produce 1.233 offspring at a rate of 1 offspring every 2.714 years (the mean inter-birth interval for females in the 8–13 year age group: see Fig. 13). This represents a reduction in ex-pected lifespan from $e_9 = 4.9$ years (i.e. at the middle of her expected reproductive career) to 0.6 years and would require an increase in the mortality rate for females from 0.204 deaths/year to 0.625/year, an increase of about 300%. This is clearly very high, much higher in fact than could reasonably be expected. (It would, of course, be partially offset by the natural pre-adult mortality of the two infants, which would be equivalent to a net loss of approximately 0.071 infants; nonetheless, the main cost still has to be the mother's own mortality.) The female's "regret" at the male's desisting from infanticide (i.e. her gain from infanticide) is equal to the number of infants gained from premature cycling, less the infant lost through infanticide and the additional mortality imposed on the new infant, a *negative* gain of −0.771 offspring. Thus females can in practice expect a net gain of 1.229 − (−0.771) = 2.000 from compromising on their reproductive strategies when a new male takes over the unit.

We have here considered only the case of a female who already has a young infant at the time of the takeover. Mori and Dunbar (1984) also reported 5 cases of females who aborted foetuses (ages ranging from ca 2 months to near term) following takeovers at Sankaber and Gich. This represents 10.6% of the females in the 7 units for which data are available. How would these females view the situation? If the mean age of the foetus at abortion was 4 months and there is an average of 2 oestrous cycles to conception, the female loses about 6 months: at the mean inter-birth interval, this is equivalent to 0.233 offspring (approximately 5% of her expected lifetime output). However, her regret on the new male's infanticidal behavior if she did not abort would be the full period of gestation (6 months) plus the time to conception (equivalent to 0.311 offspring), assuming that the risk of infanticide is 100% and that it occurs immediately after the infant's birth. (In practice, the new male might be unable to kill the infant until it has started to leave its mother to play with other infants, about 2 months or more after birth.) At minimum, the female's net gain is −0.233 − (−0.311) = 0.078, a marginal gain in favor of abortion. In addition, the energy cost increases steadily with time post-conception, the highest energy demand being due to lactation (see Portman 1970), so that the energy differential will be rather greater in favor of the premature termination of the pregnancy.

The old male expects to gain all the offspring who are at risk of infanticide (in the observed sample, an average of 3.25 offspring under the age of 12 months were present in each unit). Since he can expect a tenure as harem-holder of only about 3.6 years (see p. 166), this represents around 30% of his total lifetime reproductive output. He may also expect some gains to accrue from protecting his mature daughters, though these gains are likely to be small by comparison with his savings if the new male forgoes infanticide. His regret on the new male's forgoing infanticide is equivalent to the number of his offspring that are killed by his successor, less whatever he would have gained by taking over a new unit. This last is difficult to estimate, for we have no idea how easy it would in fact be for him to acquire another unit. Nonetheless, we can get some idea of its magnitude by considering how high the probability of his acquiring another unit would have to be in order to offset the loss of 3.25 infants by infanticide. With an average harem size at takeover of 4.71 females (Fig. 54), a male who takes over a new unit can expect to gain $0.5 \times 4.71 = 2.36$ offspring immediately through the premature cycling of half the females he acquires. The remainder ($3.25 - 2.36 = 0.90$ offspring) would have to be made up as the remaining females come into oestrus during the course of their normal reproductive cycles. At the observed mean inter-birth interval, this would take $0.90/2.143 = 0.420$ years. Since he would need to stay in the unit until the youngest of these offspring was past the risk of infanticide, he would have to retain control of the harem for a period equivalent to the delay to the last conception (0.42 years) plus the duration of gestation (0.5 years) plus the period of infanticide risk (1.0 years), a total of 1.92 years. Now, males acquire harems by takeover at the age of 7.7 years on average (Table 22) and lose them again 3.6 years later (Fig. 60) at 11.3 years of age. With an expectation of life at age 11.3 years of only a further 1.8 years (Fig. 11), a male would be unable to make up his loss even if he acquired a second unit straight-away. In fact, males have only a 70% chance of successfully taking over a unit on each attempt (Table 22), and this is for males in their prime who have not just been involved in a prolonged fight. If we assume that the male would require as little as 2 months to recover after each defeat, we can derive a rough estimate of the number of offspring he would gain if the probability of acquiring a harem of size 4.71 females is $p(A) = 0.70$, the probability of retaining it for a year is $p(S) = 0.735$ (from equation 3, Table 24), and the interval between successive attempts at takeover is 2 months. Since he has to survive as harem-holder for 1.5 years after an infant's conception to ensure that it will survive the infanticidal period, we need only consider the expected gains of males that take over a unit 2 and 4 months after losing the first one. Greater delays result in the male being unable to fit in the required period as harem-holder before he dies. The expected gain is equal to 70% of the output of a male who acquired a second harem on the first attempt plus $30\% \times 70\% = 21\%$ of the output of a male who failed on the first attempt and succeeded on the second. This turns out to be 1.555 offspring, which leaves him with a net gain of $-3.25 + 1.555 = -1.695$ offspring if he allows the new male to be infanticidal. Consequently, his overall gain if he remains in his original unit is $3.25 - (-1.695) = 4.945$ offspring. (We can ignore the loss of 0.62 infants due to abortions since this is a constant on both sides of the equation.)

The incoming male's gain is rather easier to compute. When he first joins the unit, he can normally expect all his females to come back into oestrus during the next 2.143 years. Thus, in a 3.6-year tenure, he would expect each of his females to resume oestrous cycling $3.6/2.143 = 1.680$ times. With an average of 5.27 females over his period of tenure (midway between the 4.71 he starts off with and the 5.83 he ends up with), he would produce 8.854 offspring during his reproductive lifetime. However, 50% of the females come into oestrus immediately following the takeover, and these females would all cycle again while the male was harem-holder. Thus, taking into account females who cycled prematurely, his output would be:

$$5.27 \text{ females} \times [(0.5 \times 2.680) + (0.5 \times 1.680)] = 11.489 \text{ offspring,}$$

a net gain of 2.64 offspring. His regret on desisting from infanticide is 3.25 offspring (equivalent to what he would have reduced the old male's reproductive output by), plus the number of offspring he would have gained from the females coming into oestrus prematurely as a result of his infanticidal behavior (i.e. 2.64 offspring). Thus, his overall gain from forgoing infanticide is -3.25 offspring, plus whatever gains may accrue from the help he receives from the old male in defending the unit's females. Discounting the latter, it would seem that he would actually do better by being infanticidal. Clearly, there must be good reasons why he should refrain from pursuing this strategy.

There would seem to be only three likely explanations: (1) that there is a substantial asymmetry in pay-offs, such that the old male will be prepared to fight much harder in defense of his last offspring than the new male is prepared to fight to get rid of them; (2) that, because the new male will himself eventually become an old male, it does not pay males to be infanticidal in the long run since over a lifetime they gain more by refraining from infanticide; and (3) that the new male's gain from the old male's help in defense of the unit far outweighs his loss on refraining from infanticide.

On balance, the first possibility seems implausible, if only because males of other species are quite capable of driving off their defeated predecessors and killing their offspring. The second explanation runs the risk of Alexander's parent-offspring fallacy (see Dawkins 1976) because it is susceptible to cheating. Nonetheless, although some degree of cheating is possible, it is not in the males' interests if they all cheat.

The situation thus resembles the Game Theory problem known as Prisoner's Dilemma: in this problem, the gains for both players are highest if they cooperate but lowest if they both try to cheat (see Rapoport and Chammar 1965). This game is known to have a stable solution of mutual cooperation when the series of games is infinite (as it is in the evolutionary case). It is therefore possible that, in the present case, the situation might stabilize at a low level of cheating with most males trading reciprocal altruism. The third explanation is also plausible, but it implies that the new male runs a very high risk of being himself displaced fairly soon. To break even, the new male would have to lose his unit 2.15 years earlier than he would otherwise expect to (this being the length of time it would take his 5.83 females to produce 3.25 offspring at an inter-birth interval of 2.143 years, plus the time those infants would take to pass the infanticide age threshold of 1 year). However, the old male will remain in the unit only for about 2 years anyway (see below), which makes it unlikely that an increased likelihood of losing the unit in the later part of the new male's tenure is of any relevance. In any case, there is little to suggest that the old males are of much value in preventing takeovers. Of 10 units at Sankaber which had old followers, one was taken over during a 57 unit-month period ($p = 0.018$ per unit per month), while two of 16 large units that did not have old (or young) followers were taken over during a 153 unit-month period ($p = 0.013$ per unit per month). The difference between the probabilities is much too small to be a plausible reason for the old male staying on. Nonetheless, it is possible that the old male's main value lies in preventing a further takeover during the first few months immediately following the new male's acquisition of the unit when relationships within the group are still unstable. Multiple entry-attempts are more common than expected by chance and 50% of the units that were taken over were subjected to entry-attempts by several males (Table 31). In at least 5 of the 7 takeovers, the new male had to devote a good deal of time and energy to fending off other males after he had won the unit. The new male stands to lose his entire reproductive output (11.49 offspring) if he loses the unit immediately after having acquired it; the loss on refraining from infanticide (3.25 offspring) is small by comparison. The size of the differential implies that the male's probability of himself being taken over before being able to breed would have to be very high. This can be determined by evaluating the expression:

$$3.25r = 11.49(1 - t),$$

where r is the probability of refraining from infanticide and t is the probability of premature takeover. Since $r = 1$, $t = 0.72$, an extremely high value all things considered. Since old males do not normally leave their former units immediately after takeover, there is no way in which this hypothesis can be tested without resorting to experimental manipulation. One could perhaps point to the fact that one of the 7 males who successfully took over units did suffer a partial loss of his new harem within days, despite the presence of several old males. Evaluating the relative importance of the second and third hypotheses is therefore rather difficult, though it seems that either or both could well be operative.

In summary, it seems that both the males and the female gain from the arrangement, although the new male may initially sustain a loss in order to do so.

How Long Does the Old Male Stay On?

Without the benefit of a long-term study, it is difficult to say how long the old male stays on in the unit after being taken over. All we can say for certain is that one old male, taken over in May 1971, was still with his unit in July 1975, 4 years later, while two other males (who must have been taken over sometime between April 1972 and October 1974) died while still in their units in November 1974 and May 1975, respectively. The presence of three *old* followers in unit N5 in 1975 (one known and two presumed takeovers) suggests that former harem-holders can accumulate if a unit is taken over repeatedly in quick succession. It is, however, equally clear from the composition of the all-male groups in 1974–75 that at least some former harem-holders eventually go back into an all-male group: there were two very old males (more than 12 years old) in the Sankaber all-male groups in 1974 who could not have been present as younger males in 1972, as the oldest males in the all-male groups during the first study were no more than 6 years old.

An indication of what might happen is provided by the old follower of H60. This male was taken over in May 1971, and remained with his unit right through to the end of the second study in July 1975. During the final 6 months of field work, however, he did seem to be less closely attached to his unit than he had previously been. By June 1975, he was often elsewhere in the herd, sometimes alone, sometimes near an all-male group. He never had any problems rejoining his old unit and his relationship with his successor as harem-holder had throughout been excellent. But, in the later stages, his ability to interact easily with members of other units, and especially males from the all-male groups, did begin to attract our attention. Members of other harems would reply to his contact-calling as he passed them, a behavior that is quite unusual among the gelada (unpublished data). In a nutshell, his relationships with other members of the band seemed to be much more relaxed, and he seemed much less concerned to keep up with his old harem when they moved ahead during the day's foraging. It is no doubt significant that, by this stage, the last of his female offspring were reproductively mature, while his male offspring would have left their natal unit to join an all-male group. Thus, aside from his decreasing value as a protector to the unit due to his rapidly declining physical powers, the main reason for his staying with the unit (i.e. the protection of his immature offspring) was no longer relevant.

At least two old males, however, died while still members of their units. We can use these data to obtain an estimate of the mean length of stay as an old follower. Two deaths among a total of 7 old followers in a 9-month period gives an annual rate of 0.381 per male. The reciprocal of this (2.62 years) is equivalent to the median period of residence. Counting the old follower of H60 as a loss (even though he had not quite severed the links with his old unit yet) yields a mean annual rate of loss of 0.571, which gives a median residence of 1.75 years. Since these can be considered as minimum and maximum values, the average (2.10 years) is probably a reasonable estimate. Owing to the small sample size, the 95% confidence limits around this estimate are rather wide (1.20–4.98 years); however, the upper limit is unlikely to be greater than 3.00 years because of life-historical constraints.

Is the Male Strategy-Set Evolutionarily Stable?

To be evolutionarily stable in the sense defined by Maynard Smith (1974; Maynard Smith and Price 1973), a set of strategies must each yield the same net contribution to the species' gene pool when their frequencies of occurrence are taken into account. We know from the analyses in Chapter 12 that the two main elements in the gelada male's strategy-set do yield the same number of offspring over a lifetime. Nonetheless, it need not be the case that the genetic fitnesses of the two strategies are the same: if the strategies differ significantly in frequency, the contributions made to the species' gene pool by the two classes of male would be radically different and the strategy-set itself could not be in evolutionary balance. In order to determine whether the strategies are in equilibrium, we need to know the relative frequencies of the two strategies. In addition, we also need to know not

Table 46

Frequencies with which adult and subadult males entered reproductive units in relation to the numbers of males in the population that had not yet started their reproductive careers.

	Adult	Subadult
Males entering units as followers (n/yr)	6.7	5.3
Males taking over units (n/yr)	6.7	0.0
Total males entering units (n/yr)	13.4	5.3
Males available in population[a]	11	58

Source of data: Main and Abyss bands in 1971–72 and Main band in 1974–75; all entry rates are corrected to give annual rates.

[a] Young followers and all old males excluded.

just how many males succeed in entering units by each of the two strategies, but also whether any males actually fail to acquire harems during their lives.

Do All Males Get Harems Eventually?

Without the benefit of a 20-year study, we cannot hope to determine whether every male gets a chance to own a harem at some point during his lifetime, or whether in fact some males remain confined to all-male groups and never breed. However, some indication can be obtained by considering the observed rate at which males entered units in relation to the number of males who had not yet begun their reproductive careers. Table 46 gives the number of males entering reproductive units, either as followers or by takeover, during the two field studies at San-kaber, together with the number of males in the study populations who were not harem-holders or followers at the start of each of the studies. Old males were excluded from this analysis on the assumption that they had already had their turn. The data suggest that, over the course of a year, slightly more than 13 adult males and 5 subadult males embarked on their breeding careers by entering a reproductive unit. The sample population from which these were drawn contained only 11 adult males and 58 subadults.

In other words, all adult males can expect to start their breeding careers within a year, on average. Since some adult males enter more or less as soon as they mature at age 6 years, most males can expect to have entered a unit by the age of 8 years at the latest. This agrees well with the actual data, which suggest that nearly half the males enter units as followers at the age of 6 years and most of the rest take units over at 8 years of age (Table 44).

Note that very few subadults can expect to join a reproductive unit (only about 10% per year). It seems as though most males wait until they are physically mature before deciding to make their break. This suggests that males try to keep their options open rather than commit themselves to a follower strategy while still in the subadult stage of their life-cycle (cf. Rule 2, p. 136).

Thus, it seems very likely that, at least at Sankaber, all males do in the end get the chance to become harem-holders at some point during the course of their lives.

Evolutionary Equilibrium?

Having established that all males do eventually breed, we can use the data on the frequencies of the two strategies (Table 21) and the results of the simulation in Chapter 12 (Fig. 58) to determine the net genetic contributions of the two strategies. It is easy to see that the two main strategies are likely to be in evolutionary equilibrium, as, not only do they yield similar pay-offs over a lifetime, but their frequencies are approximately the same. Moreover, the decision analysis presented in Chapter 14 suggests that the net pay-offs for the two strategies do equilibrate when all possible outcomes and their frequencies are taken into account (Table 44).

This implies that the strategies are *bona fide* alternatives in the evolutionary sense and not simply instances of either an evolutionary "best of a bad job" or a developmental process (see Dunbar 1982a).

Although the strategy-set is in overall evolutionary balance, the precise ratios of the two strategies vary from year to year (and band to band) in response to changes in the demographic structure of the bands. These ratios are set by density- and demography-dependent processes determined by a band's demographic structure at any given time, but are probably held in local equilibrium by frequency-dependent feedback effects.

Several lines of evidence attest to the importance of demographic variables in determining the relative frequencies of the two strategies. In the first place, the relative frequency of takeover entries is a linear function of mean harem size (Figs. 61 and 62). Although it is easier for males to effect takeovers when the mean harem size increases (and males therefore pursue this strategy more frequently), their expectation of lifetime reproductive output becomes less than that for followers once the harem size at takeover exceeds 7.5 females, even though the follower's initial harem size remains unchanged at 1.9 females (see Fig. 58). In the second place, the larger the harem size, the more likely it is that a multiple takeover will occur (Table 31); in this event, the females will be divided among two or more males and the unit will undergo fission into a set of smaller units. Mean harem size thus tends to revert to a constant size by a process analogous to stabilizing selection. This suggests that there is a tendency for the system to stabilize at around the observed frequencies because too great a shift in favor of takeovers is counterproductive for takeover males. Note that this has nothing to do with the relative frequencies of the two strategies; it is a consequence of demographic processes and their effects on the likelihood of takeover. The same result would pertain if no males ever pursued the follower strategy.

With the optimal frequencies of the two strategies so determined, the distribution may be stabilized by one of two mechanisms. First, frequency-dependence alone may reduce the expected gains of takeover males if the likelihood of takeover at a given harem size is affected by the number of males attempting takeovers, thereby reducing the expected tenure of a takeover male (see Table 26). If the size-specific takeover rate is invariant, however, there will be an upper limit on the proportion of males who can successfully acquire harems by takeover: because there is a lower limit on the size of harem that can be taken over, only a fixed number of males will be able to take harems over at any given time and the rest will be obliged to settle for follower-entry or a further delayed attempt.

Although the two processes yield diametrically opposite predictions about the stability of the size-specific takeover rates, it is not easy to test between them since data on takeover rates in relation to the frequency of takeover entry-attempts are not available for a sample of populations with the same distribution of harem sizes. One way round this problem is to standardize the takeover rates for the various bands given in Figure 61, and then to see whether the rates are constant when plotted against the frequency of takeover entry-attempts (given in Fig. 62). In effect, we assume that the takeover rates *are* invariant and we predict that, once the rate for each band has been standardized to remove the effects of differences in mean harem size, they should be roughly constant with respect to the variation in the proportion of takeover entry-attempts. The takeover rates can be standardized by using the regression equation from Figure 61 to convert the observed rates into ones equivalent to the mean harem size for the sample (5.01 reproductive females). Figure 75 shows that there is in fact a strong linear relationship between the standardized takeover rate and the proportion

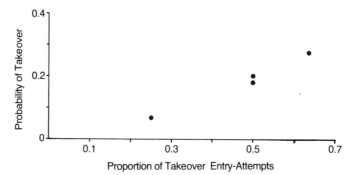

Figure 75 Standardized probability of takeover per year, plotted against the proportion of takeover entry-attempts for 3 bands at Sankaber and one band at Gich. The takeover rate is standardized to a constant mean harem size (the sample mean of 5.01 reproductive females) to remove the effect of differences in harem size distributions between the bands. Based on the data given in Figs. 61 and 62.

of takeover entry-attempts (least-squares linear regression: $r^2 = 0.988$). More importantly, the slope is significantly greater than zero ($t_2[b = 0] = 12.697$, $p = 0.012$ 2-tailed): high frequencies of takeover entries are associated with higher than average takeover rates, while lower than average takeover rates occur when the proportion of takeover entry-attempts is low. This is unequivocal evidence in support of the hypothesis that the relative frequencies of the two strategies are stabilized at their locally optimal values by frequency-dependent processes. These act in such a way as to increase the slope of the size-specific takeover rates in proportion to the pressure exerted by the relative frequency of takeover entry-attempts, this in turn being partly a consequence of the ratio of acquirable units to males trying to pursue a takeover strategy.

16 Evolutionary Decisions Under Conflicts of Interest

I have tried to show that the social behavior of gelada baboons can best be understood in terms of reproductive strategies, that these strategies involve decisions on the part of the individual, and that these decisions in turn are made in the context of a variety of constraints imposed by the biological and social environment within which the animal is embedded. I have taken a deliberately broad perspective, arguing that, in order to understand one aspect of an animal's behavior (namely, its reproductive strategies), it is necessary to understand in detail the structure of the society within which the animal lives and the ecological, demographic, and physiological constraints that act on it.

In this chapter, I reconsider some of the problems encountered in this attempt at a holistic analysis of a complex social system in order to highlight the methodological lessons we can learn from it.

Decision-Making Under Constraint

I have tended to assume that animals ultimately seek to maximize their reproductive output over a lifetime. In doing so, I ignore the fact that this in itself is only an intermediate goal in the proper Darwinian objective of maximizing contribution to the species' gene pool, an objective that since Hamilton's (1964) seminal papers we now realize can be achieved either by maximizing personal reproductive output or by maximizing the reproductive output of relatives (or by some optimal combination of the two). Nonetheless, although the kin selection component is undoubtedly important, I do not believe that it is a general panacea for all our problems and I have tried to show that other factors may be at work even where kin selection has been an obvious explanatory candidate.

This is not to suggest that kin selection is irrelevant. Indeed, we have seen that kin selection does contribute something to the reproductive value of particular strategies (notably in the formation of female coalitions). Nonetheless, what it contributes is not of such a magnitude that we would want to identify kin selection as the primary evolutionary explanation. This underlines an important methodological point. In practice, sociobiological arguments commonly rest on a single test based on a comparison of observed data with a distribution predicted by the variable of interest. In the present case, we might have wanted to assess the importance of kin selection as an explanation for coalition formation: we would have evaluated the gains due to kin selection, observed that they agreed at least on an ordinal scale with the theoretical predictions, and concluded that kin selection was indeed the correct explanation for the phenomenon. But this is only true because we have not *simultaneously* evaluated the importance of alternative competing hypotheses. Having in fact done so, we found that considerations of personal reproductive output outweigh the undoubted gains from kin selection. In general, the costs associated with kin selection are such that it can only be a major selective force when either the costs of altruism (in the genetic sense) are negligible or the likelihood of personal reproduction is very small (Dunbar 1983a).

It has become fashionable to view parental investment as a form (some would say, the *ultimate* form) of altruism (see Wilson 1975, Maynard Smith 1977, Wallace 1980). This might lead one to argue that considering only birth rates (as we have in effect done) ignores the significance of kin selection through the altruistic component of parental behavior and its influence on the numbers of offspring reared to maturity (not to mention their future breeding efficiency: see, for example, Fuchs 1982)). This argument, as I have pointed out elsewhere (Dunbar 1982d; see also Bertram 1982), is fallacious because it confuses proximate and ultimate costs and gains: parental investment is a decision about how best to maximize one's own personal (i.e. direct) contribution to the species' gene pool, not a decision about whether or not to invest in another individual's contribution. This last is a logically prior decision and often needs to be resolved *before* a decision is made on how much parental investment should be meted out (see Dunbar 1983a).

Nonetheless, parental care invested in the developing young is an important means whereby an individual can influence its genetic fitness.*

* I use the term *fitness* loosely here, since strictly speaking, fitness is a characteristic of genes (or of genetic traits), not of individuals (see Dunbar 1982b). Nonetheless, there is a sense in which we can usefully speak of an *individual's* "fitness" in terms of the numbers of copies of any arbitrary gene that the individual contributes to future generations (see also Calow and Townsend 1981).

I have ignored it because its consequences are of such a long-term nature that we could not have observed the effects of differences in parental care. Nonetheless, it was possible to show that females do differ in the extent to which they invest in their offspring. Clearly, there is, at least in principle, scope for low-ranking females to offset the disadvantages of low reproductive rates, though the sample of 10 infants is too small and the nature of developmental pathways too complex and protracted to allow this area to be explored in any detail at the present time. On the other hand, parental care does not seem to feature to any significant extent in the behavior of harem-holding males. Males seem, rather, to capitalize on the females' parental behavior in a way that affects all males similarly. The fact that we can generate a cogent, self-contained explanation of male behavior without reference to parental investment implies that its importance to males is not sufficient to allow them to use it as a means of generating alternative strategies of reproduction.

Despite this omission, it is clear that the situation is complex enough. Both females and males have a variety of options open to them at the strategic and tactical levels. In both cases, this variety of choice is the result of feedback effects, often of the frequency-dependent kind, that make the constraint-free strategy less attractive as the costs of pursuing it increase due to increasing competition.

Differences in reproductive rate between individual females are determined mainly by the females' dominance ranks, because harrassment by more dominant females suppresses reproductive activity in low-ranking females. It is this physiological effect that sets up the conditions favorable for the pursuit of alternative strategies aimed at offsetting the reproductive disadvantages of low rank. Females have two main options: forming alliances to bolster their dominance ranks or transferring to a male with fewer females.

Although females have a variety of potential allies available to them, they are not all equally valuable. The male, although a useful short-term ally in many respects, is likely to die just at the point at which his aid as a coalition partner would be most critical to the female (i.e. as her natural rank-determining abilities begin to decline with age). There are, on the other hand, considerable advantages in terms of long-term stability (and, to a lesser extent, of kin selection) to forming coalitions with close female relatives. Unrelated females (irrespective of their rank) are too likely to desert an ally in favor of a daughter when one matures. Consequently, while an unrelated female might be considered a useful temporary ally as long as neither female has any mature daughters, the fact that alliances are asymmetric and (in the short run) non-reciprocal makes the likelihood of lost investment too high to consider it a worthwhile option. This doubtless explains why females who have no mature close female relatives in their unit make little effort to form coalitions with other females.

The alternative option of deserting her male for another with a smaller harem is subject to some (as yet unidentified) constraints on female mobility. The gelada contrast with the hamadryas in this respect, for hamadryas females are able to move relatively easily from one unit to another (Sigg et al. 1982). This implies that the selection pressure against doing so in the gelada is severe, for its advantages as a reproductive strategy would appear to be considerable.

Since female reproductive rates tend to be more constrained than those of males, females tend to be more selective than males in their behavioral strategies; they thus exert proportionately more control over the structure of the social system (Goss-Custard et al. 1972, Wrangham 1981). This has important implications, for it means that males will not necessarily be able to pursue the strategy that, other things being equal, would give the best return in terms of reproductive output. Rather, whenever the males' interests conflict with those of the females, then males will be forced to compromise on their reproductive strategies. Thus, males would presumably prefer to hold the largest possible harems (subject only to the limitations imposed by the ecological densities at which females could subsist), but it is not generally in the females' interests to live in such large groups because of the reproductive costs that group-living imposes on them.

A female's loyalty to her male seems to depend mainly on how well she is doing at the time. Hence, the male's problem is essentially one of trying to balance the gains to be derived from large harems against their costs in terms of the difficulty of retaining large units for any length of time. These costs are a consequence of the females' strategic preferences, although the actual rates of takeover in any given case depend on the demographic structure of the population. This, in turn, is determined by the interaction of environmental conditions, demographic processes, and social factors. Between them, these determine the two key variables that affect takeover rates: the distribution of harem sizes and the number of males waiting for the opportunity to own a harem.

While we have been able to undertake a detailed analysis of male reproductive strategies and to demonstrate the existence of clear-cut alternatives that equilibrate reproductive output over a lifetime, we have not been able to do this with the same success for the females. This is mainly because female strategies are much more fluid and less determined than those of males. A male makes an early decision about his reproductive career, and he is then committed to that decision with little prospect of being able to make other than minor tactical adjustments to offset a bad strategic choice. Females, on the other hand, make their reproductive decisions in a situation that is open to more random effects (notably, the presence or absence of female offspring and sibs), offers more options for continuously responding to circumstances at both a tactical and a strategic level, and operates in an altogether much subtler and less dramatic way than is the case for males. This makes it difficult to define a clear set of alternative strategies whose lifetime outputs can be evaluated: a female can pass through alternating periods in her life during which she may, and then may not, have female relatives with whom to form alliances. No doubt this is why there is so much variation in the numbers of offspring produced by females of different ranks, with some low-ranking females having more offspring than some high-ranking females (see Dunbar 1980b, Fig. 6). Consequently, despite the importance of dominance as a determinant of female reproductive rates in the short-term, the lifetime reproductive outputs of females of different rank appear to be more or less equivalent (Fig. 23).

The Study of Inherently Complex Systems

It is important to bear in mind that most of the analyses in this book are specific to one population of gelada. It would be naive to assume that they apply to other species, since the constraints that act on each species differ in significant ways. However, to conclude from this that there are no universal principles in behavioral biology (i.e. principles of the same logical status as the laws of physics or chemistry) is to confuse different levels of explanation. Probably the only principles that will be universally valid lie at the genetic level (or perhaps at the sub-genetic level of lifetime reproductive output): the rule may be quite simple (e.g. "Maximize contribution to the species' gene pool"). The way in which such an objective is achieved in any given case will depend on the specific ecological and demographic contexts in which the individuals concerned make their reproductive decisions. These will vary not only from species to species, but also between populations of the same species (or even the same population at different times).

Nonetheless, we can still identify universal principles at this "lower" level, but they are universal only in the conceptual, not the observable, sense. That is to say, for any given behavioral problem, there will be an optimal solution which will in fact have the status of a universal principle: it will always be valid only providing this component of the biological system is taken in isolation (a restriction that is in fact also true of most physical laws). Other things being seldom equal in practice, however, the various components of the system will tend to be in competition and the final solution for any given problem will depend on how the set of problems (and the universal principles that specify their individual best solutions) interact. The problem is, thus, one of optimizing, not maximizing, output, and the final choice of strategy will necessarily be context-specific and thus a compromise. To look for a universal rule that is valid in all circumstances is to misconstrue the nature of both biological systems and universal laws.

The immediate problem raised by this view is whether it is even possible to set about testing theoretical models (or, indeed, more qualitative *a priori* hypotheses). The short answer is that it can only be done providing the system as a whole is sufficiently well understood for all its components to have been included in the model. Only then will it be possible to say with any certainty that a prediction made about another population (or even the same population at a later time) will not run foul of some unstated *ceteris paribus* clause. The essence of the problem is that, whenever a prediction is not upheld by the data, we do not know whether the model itself is conceptually incorrect or whether the model is correct as far as it goes, but all other things were not equal because a key element in the system had been omitted.

In most cases, the only practicable solution to this problem is a combination of comparative intraspecific studies and the building and analysis of multifactorial models based on these studies. In the short run, of course, there is still much that can be done by studying specific problems in isolation. But this is mere tinkering and in the long run we will have to come to grips with the integrated nature of biological systems if we are ever really to understand social behavior. I hope that the present analyses go some way to justifying this view.

This raises one further problem: given the complexity of the system and the difficulties of obtaining large samples of data on each of many relevant variables, how reliable are such analyses likely to be? Inevitably, many of the estimates of parameter values will be based on relatively small samples, and these will tend to prompt questions about small sample biases. The optimization approach to the problems of social behavior advocated by sociobiology differs from the more traditional ethological orientation in one key respect: many of the hypotheses under test are not null hypotheses generated by assumptions of random interaction from which we seek significant differences, but quantitative predictions to which we seek similarities. Davies (1982) makes a similar point with particular reference to ESS analyses, for these invariably predict equality of pay-offs. Since, conventionally, we test for non-significant differences rather than for significant similarities, type II errors (accepting the null hypothesis of no difference when it is in fact false) become increasingly likely as the sample size gets smaller. This is undoubtedly a problem, but it is, I believe, no more serious a problem than the converse one of ignoring large parts of a coadapted system in order to study a single component of it in great detail. It is certainly not a problem that we should dismiss as insoluble, thereby rendering holistic studies worthless. Rather, it should be seen as a challenge to our ingenuity as scientists.

Insofar as the present study is concerned, there are three main reasons for considering this worry to be less serious than it might at first sight seem.

In the first place, a great many of the sample sizes are far from small by the normal standards of either field or laboratory research: between them, the Simen studies involve more than 1500 animal-*years* of observation. This inevitably reduces the overall variance in the estimates for an individual data-point. Even though a regression may be fitted to points for only 3 or 4 habitats, the large samples mean that each point is estimated with minimum error. Only a few variables are based on genuinely small samples. Where the rates with which specific behavioral events occur have been especially low, I have tried to reduce the variance in the estimates either by applying standard smoothing techniques or by using the average of estimates obtained in different ways.

The second reason is that throughout the analyses we have been exceptionally fortunate in having independent samples against which to test hypotheses, notably in the data available from Gich. In addition, much of what we have been able to achieve in this analysis of gelada behavior is a testimony to the value of comparative studies. These have not only provided us with the basis for determining *quantitative* relationships between variables, but have also helped to explain many aspects of the behavior of the Sankaber gelada. A number of results that were originally put down to inherent error or biased sampling have turned out to be genuine. As just one example, consider the finding that there were fewer immatures per unit in Gich bands compared to Sankaber in our 1971 data, despite similar harem sizes. This led us to conclude (Dunbar and Dunbar 1975, pp. 130–131) that the difference in the numbers of immature animals in the units was the result of biased sampling over too short a time at Gich, and so to caution against placing too much weight on data from such brief studies. In fact, the difference turns out to be genuine, being a consequence of the influence of climate on reproductive rates (Ohsawa and Dunbar 1984).

The third consideration is simply one of parsimony. The whole story hangs together much too well to be the coincidental consequence of a number of unrelated small sample biases. Caution is undoubtedly justified when a single result is at stake, but as the number of interdependent results increases, the validity of the explanation linking them increases correspondingly since complex explanatory chains supported by some data at key points are proportionately less likely to arise by chance. To suppose otherwise would require a monumentally complex explanation in terms of coincidences that would become increasingly unlikely as coincidence was added to coincidence.

Evolution and the Problem of Consciousness

It is not my intention here to assess the general problem of consciousness in animals, even though it is becoming of increasing interest to ethologists (see Humphrey 1976, Griffin 1981, Crook 1980, 1983, Kummer 1982). I shall simply take it as axiomatic that animals make decisions (where by *decisions* I mean non-random choices based on an assessment of the context and the options available). The more pressing problem is how these decision processes relate to evolutionary explanations of behavior.

The question of decision-making in strategy choice by animals necessarily forces a re-evaluation of the traditional dictum that terms describing human conscious processes should be avoided at all costs in discussions of animal behavior. At its worst, this view gave rise to the ultimately sterile operationalism of the Behaviorist school in the 1930s. On a more general level, however, its continued acceptance may be counterproductive. Kummer (1982) stresses this in his thoughtful discussion of the problem of social knowledge in primates. He points out that the oversimplifications generated by too close an adherence to Lloyd Morgan's canon have a number of undesirable consequences. These include (1) an artificial widening of the differences between animals and humans (the kind of muddled logic that argues "I know how humans think, so they must be conscious; I can't be sure how animals think, so they can't be conscious": see Dunbar 1982e); (2) making our subject animals appear more simple than they really are, thereby restricting the investigation of behavior to low-level hypotheses of the simple stimulus-response kind; and (3) diverting attention away from the more complex aspects of sociality. In recent years, studies of higher primates in particular have been forcing ethologists away from the traditional species-specific view of behavior to a more sophisticated view of animal behavior in terms of decision-making and strategy-assessment (see Kummer et al. 1978, Walker Leonard 1979, Stolba 1979, Strum 1982). We have become increasingly aware that animals can compensate for physical disadvantages by resorting to relatively sophisticated social stratagems (see, for example, Anderson and Mason 1978). Similarly, it is only recently that we have become aware of just how sophisticated the communication systems of primates (in particular) are in the extent to which they can convey syntactical information (see Snowdon et al. 1982).

My contention is that we cannot study these problems adequately without resort to the language of assessment and decision. The "objective" language of neo-behaviorism is just too restrictive; it imposes an artificial limitation on an animal's behavior at a conceptual level. This does not mean that we are *ipso facto* obliged to attribute to animals the full complexity of the conscious processes that humans are capable of; it is merely to admit that the difference is one of quantity rather than quality. (It is perhaps relevant that many of those who decry this view either work with the more primitive taxa where many of these problems clearly do not arise, or have never studied animals in their natural habitats.) Nor does it mean that we are obliged to suppose that the animals assess the value of the strategies they pursue in terms of the units of genetic fitness (or even numbers of offspring) that they will provide over a lifetime. As McFarland and Houston (1981) have pointed out, there will often be a clear distinction between the ultimate costing by which strategies are assessed on an evolutionary scale and the rules of thumb that the animals themselves use in their decisions. The latter will, however, have evolved precisely because they correlate tolerably well (or at least better than anything else does) with genetic fitness.

This is not just a plea for tolerance in epistemological matters (*sensu* Feyerabend 1963). There is a crucial methodological point at stake: to be able to study these kinds of social systems, the observer has to second-guess what his animals are up to, and this means thinking himself into the animal's "state of mind" in just the same way as historians do for past events and cultures (see Gardiner 1952). Unless this can be done, it will often not be possible to appreciate the options that are available to an animal, nor to understand how its biological capacities limit these options and constrain its behavior. This step is essential precisely because we need to determine not merely what an animal does, but, more importantly, what it is *trying* to do, in order to be able to understand why it ends up doing what it does by way of compromise (see also Seyfarth 1980). Assuming "consciousness" may in fact be a heuristically valuable procedure, at least with the taxonomically more advanced species (see also Geach 1975). Although this assumption may place a greater onus on the abilities of the individual researcher if anthropomorphic fabrications are to be avoided, in practice it may do no more than formalize what every good ethologist in fact already does intuitively. In any case, the failings of some practitioners of an art do not count as a rational criterion by which to judge the general validity of a methodological perspective (a point that seems to have escaped some of the more ardent critics of the sociobiological program, just as it did the earlier critics of Darwin himself).

This much having been said, we are at once confronted by the problem of how decision-making in animals relates to sociobiological models of the evolution of behavior. If a decision is not genetically programmed, can it still be sociobiologically interesting?

The answer, of course, is that decision-making is precisely what sociobiology is all about, and we should not be sidetracked into supposing that sociobiological explanations necessarily demand genetic determinism. They demand only that, ultimately, gene frequencies be influenced: there is no *a priori* reason why the genes whose frequencies are affected should necessarily be those directly responsible for determining the particular choice of strategy.

There is, however, a more serious aspect to this, for an inherent geneticism underlies the notion of an evolutionarily stable strategy (or ESS). This is not surprising, given its conceptual origins in population genetics. In fact, the defining characteristics of an ESS have come to lie primarily in the frequency-dependence of costs and benefits and the equilibration of final pay-offs, although in reality this is only one of several mechanisms for achieving evolutionary stability. It is essentially an optimization approach and, stemming as it does from Game Theory, it has little to do with genes *per se*, as a number of authors (e.g. Parker 1978, Maynard Smith 1978, Davies 1982) have in fact appreciated. The question of genetic determinism remains a strictly optional issue: providing that the behavioral strategies ultimately influence genetic fitness, they will reach some optimal distribution that is "evolutionarily stable." To demand that ESSs be genetic strategies (as does Dawkins [1980]) is to put an unnecessarily strong restriction on the use of a heuristically valuable approach. Calow and Townsend (1981) come to much the same conclusion in their lucid analysis of the structure of evolutionary explanations.

A more serious problem that does need to be considered is the question of precisely what is being equilibrated. A strict genetic determinism will demand that, since the ultimate costing is inevitably in terms of future gene frequencies, then it is genetic fitness that will be equilibrated. Consequently, the lifetime outputs of individuals pursuing different strategies could be radically different. I would argue that this is possible only if the animals concerned have no control whatsoever over their choice of strategy. Once animals are capable of influencing their own reproductive outputs by making decisions (however limited) about their behavior, then evolutionary processes will tend to shift the level of equilibration to those more immediate levels where the more flexible decisions of individual animals could be expected to have a more direct influence. Towsend and Hughes (1981) provide a nice illustration of this with respect to great tits and caddisfly larvae: these, respectively, can and cannot make such an adjustment and so, respectively, can and cannot forage in a locally optimal way. This shift of the level of equilibration is inevitable precisely because informed phenotypic adjustment of behavior will almost always yield higher fitnesses in the end than genetically predetermined responses. Parker and Rubenstein (1981) have shown that, if the organism's neurological capacities are such as to allow the ability to assess to evolve, it should invariably do so. This may seem trivially obvious to many ethologists, but the point needs emphasis because an overly genetic orientation to sociobiological problems seems to be in danger of leading us into a study of population genetics rather than a study of animal behavior. By the same token, it also tends to discourage the wider acceptance of a sociobiological perspective in the behavioral sciences.

The final problem lies in a hidden feature of ESSs and concerns the basis on which an individual actually makes its decision. In the classic case, a frequency-dependent ESS occurs because the participants in a contest have no prior knowledge about the capabilities of their opponents. The decision about which strategy to adopt was viewed as a genetic "decision": because the choice is made at conception, there is obviously no way in which the fertilized egg can predict whom it is likely to meet during the course of its future life. Subsequently, this view was modified to take account of the information that a contestant could acquire about an opponent (notably via displays) and led to the idea of a "conditional ESS" in which relatively simple rules of thumb dictate what an individual should do under a restricted range of conditions (Davies 1982). Information allows an individual to select the locally most efficient strategy within the constraints imposed by its own abilities. In either case, frequency-dependence adjusts the relative frequencies of the alternative options, but only in the first case (i.e. no information) is frequency-dependence alone the basis on which each individual makes its particular choice of strategy.

In practice, instances in which this condition of uncertainty (or no information) holds are likely to be rare (see Dunbar 1982a). The gelada provide an interesting case in which it does, for males have no basis on which to predict the future demographic structure of the population (see Chapter 14). Not only does the ease with which they can take a unit over depend on the distribution of harem sizes at the time they make their attempt, but it also depends both on how many other males will be waiting for the opportunity to hold a harem and on what proportion of these have (or will) opt for a follower-entry instead. The only rational solution is to opt for each of the two strategies with equal probability, since each strategy has approximately the same expectation of reproductive success. This is precisely what the males do.

But do they really choose which strategy to pursue at random?

There seems little reason to suppose that the males actually do so, for the ages at which they make their initial attempts barely overlap (see Fig. 56). This implies some degree of prior determination of strategy choice, but we have no *a priori* grounds for deciding whether this commitment is the result of a genetic predisposition or whether it is the result of a phenotypically determined strategic decision by the animal. Two lines of evidence incline us towards the latter explanation. These are (1) the fact that, as a male's age increases, he tends to take over increasingly larger harems in an apparent effort to compensate for his reduced expectation of tenure (Table 30) and (2) the fact that some males first gain entry as a follower, only to take the unit over later (see p. 171).

The males do not, however, make "free" choices: their decisions are constrained both by the circumstances in which they find themselves and by the unique life-histories that they bring with them. These constraints are suggested by two facts: (1) the sizes of the harems the two types of male attempt to enter barely overlap (Fig. 51) and (2) males do not seem to leave their options open very long (Chapter 11, p. 136, Rule 2). While it is likely that a male who is unusually large or aggressive for his age might opt for a future takeover, these are traits that are subject to considerable phenotypic adjustment during the course of an individual's developmental history (see Dunbar 1982b). A male will therefore make his decision on these bases relative to his assessment of his standing in the local population at the time. Indeed, there is evidence to suggest that not every male's decision is optimal: older males who opt for follower-entry (Fig. 56) are almost certainly pursuing a "best of a bad job" solution, for they cannot do as well as they would otherwise have done had they got in as followers earlier on or taken over a larger unit at their current age (Fig. 58). The suspicion is that such males, having passed over the option of an early follower-entry in favor of an attempt at a more profitable later takeover, have subsequently found their options on takeover foreclosed by a shortage of units that they as individuals could successfully take over.

It is likely that there is no simple unitary explanation for the behavior of all individuals. The most plausible interpretation of what is happening is that the overall frequencies of the two strategies are determined by demographic and life-history processes, while the actual distributions are stabilized at these values by frequency-dependent considerations acting on the relative costs and benefits of the two strategies under those conditions (see Dunbar 1982a). Within this constraint, each male probably makes his own decision opportunistically in the light of his unique life-history and the circumstances in which he happens to find himself at the time. As a result, different males may be making decisions of quite different evolutionary status.

Appendices: Outline of Computer Programs

A. Female Reproductive Strategies Model

A.1. Strategy-Specific Rank Trajectory

The lifetime rank trajectory of a female was determined by assessing her probable rank in each successive year in a harem of fixed initial composition, assuming that

(a) all females enter the adult cohort at age 4 years and die at 14 years of age;

(b) a female's oldest daughter matures into the adult hierarchy when the mother is 9.5 years old, and successive daughters mature at 3.5 year intervals (see below);

(c) females rank relative to each other in the order young adult (6–8 years), old adult (9–10 years), subadult (4–5 years) and very old adult (11–13 years); and

(d) within these age classes, females rank in ascending order of age for subadult and young adult females (i.e. oldest highest) and descending order of age for old and very old females (i.e. youngest highest).

The harem composition at the start of the subject's reproductive career was chosen so as to mirror the average unit in the population (4 reproductive females, comprising a subadult, 2 young adults and an old female). The exact ages of the females depended on the birth rank of the subject, but were determined on the assumption that all the females were related to each other as the living descendants of a recently deceased female. Their relative ages, the timing of the maturation of females who were immature at the start, and the timing of successive births were determined by using constant mean age differences. These were calculated from the observed birth rate (inter-birth interval of 2.143 years), assuming a 50:50 sex ratio at birth, as follows.

The mean age of a female at the birth of her first daughter is given by

$$B + \sum_{i = 1}^{k} 0.5^{i}[2.143(i - 1)],$$

where B is the age at first reproduction (in this case, 4 years), i is the birth rank of an offspring, and k is the limit imposed on i by the female's longevity (with the observed mean age at death 13.8 years, $k = 5$). This turns out to be 5.74 years. The delay to the birth of the female's *next* daughter can be obtained, analogously, by evaluating the expression

$$\sum_{i = 1}^{k} 0.5^{i}(2.143i),$$

where $k = 4$. This is also the mean age gap between successive sisters and turns out to be 3.48 years. For graphical convenience, these values have been taken to be 5.5 and 3.5 years, respectively.

The harem's composition in successive years was determined iteratively from its initial state by applying rules (a) to (d) above.

The case for the average female is illustrated in Table A.1. The mean rank of the female in each year of her reproductive life is given in the two right-hand columns for the conditions of no coalition formation and coalitions formed with mother and daughter(s). The changing pattern of harem structure given in Table A.1 probably mimics the sequence experienced by the average female during her lifetime fairly closely. Individual females will, of course, experience a wide range of initial compositions.

The average female is equivalent to the third-born offspring. The initial composition in her case was taken to be females aged 4, 7.5, 12, and 7 years (the subject, her older sister, her mother, and her mother's younger sister, respectively: see Table A.1). The compositions of the harem when a first-born and fifth-born offspring matures were taken to be sets of females aged 4, 8, 12, and 7 years (subject, her mother, the mother's oldest sister, and this female's oldest daughter, respectively) and 4, 7.5, 11, and 5.5 years (subject, her older and younger sisters, and the subject's niece, respectively).

A.2. Strategy-Specific Lifetime Reproductive Output

The following generalized algorithm was used to determine the gain over the constraint-free strategy of no coalition formation that

Table A.1

Relative ages of the females in a harem during the reproductive life of the average female, together with her rank in each year, assuming that she formed coalitions with her mother and daughters. Her age-specific rank under the constraint-free condition of no coalition formation is given in the final column. (See text for details.)

				Females' Ages (years)									Subject's Age-Specific Rank	
D1[a]	Subject	D2	GN	N1	S1	Mother	S2	N2	SC	C1	Aunt	C2	Coalition	Non-coalition
	4				7.5	12					7		3	3
	5				8.5	13					8		3	3
	6			4	9.5		3.5			3.5	9		1	1
	7			5	10.5		4.5			4.5	10		1	1
	8			6	11.5		5.5			5.5	11		1	1
3.5	9			7	12.5		6.5			6.5	12		3	3
4.5	10			8	13.5		7.5			7.5	13	4	4	4
5.5	11		3.5	9			8.5			8.5		5	6.5	6.5
6.5	12		4.5	10			9.5		4	9.5		6	4	8
7.5	13	4	5.5	11				4	5	10.5		7	2	10

Note: Arrows indicate mother-daughter relationships.

[a] D = daughters; S = sisters; N = nieces; GN = grand-niece; C = cousins; SC = second-cousin.

would be obtained by forming a coalition with another individual during a specified period of a female's reproductive life.

1. Calculate the number of offspring produced over a lifetime through coalition formation during a specified period of that lifetime:
    (a) for each year in the female's life, determine whether or not she is in a coalition;
    (b) determine her age-specific rank (from the appropriate rank trajectory determined as above);
    (c) determine her rank-specific birth rate (from [1b] using the equation for Table 6);
    (d) adjust for her age-specific fecundity (from [1c] using the equation for Figure 15, relative to a baseline equivalent to a female of median age [8.5 years old]);
    (e) cumulate total reproductive output to current year;
    (f) repeat from [1a] for the next year, until the mean age at death (14 years).
2. Calculate the corresponding number of offspring produced by a female who does not form coalitions (by applying steps [1a] to [1f] to the appropriate rank trajectory).
3. Determine the gain in reproductive output obtained by forming coalitions (by subtracting [2] from [1]).

B. Male Reproductive Strategies Model

1. Read in age structure of female complement of unit at start of tenure.
2. Calculate expected number of offspring produced by males over a lifetime with the specified harem composition (given in [1]) for each age at the start of tenure from 6 to 12 years:
    (a) Calculate number of female offspring produced in current year by
        (i) calculating age-specific female birth rate for each female year class (from equation 1, Table 24);
        (ii) calculating number of female offspring born to each year class (from [1] and [2a.i]);
        (iii) calculating gross output of female offspring in current year (summing [2a.ii] across all year classes);
        (iv) devaluing [2a.iii] by harem size effect (using equation 2, Table 24, relative to a baseline equivalent to a harem of 4 females).

   (b) Calculate male's total output in current year by
      (i) calculating male's likelihood of not being taken over in current year (from equation 3, Table 24, given current harem size from [1] adjusted by [2d]);
      (ii) calculating male's probability of survivorship to midyear (from equation 4, Table 24);
      (iii) computing conjoint probability of retaining tenure during current year (from [2b.i] and [2b.ii]);
      (iv) calculating number of offspring of both sexes contributed to lifetime output by males surviving to current year (by devaluing [2a.iv] by [2b.iii] and doubling, assuming a neonatal sex ratio of 50:50).
   (c) Cumulate net reproductive output to current year.
   (d) Recalculate female composition of harem for next cycle (year) by
      (i) setting number of females in 0th age class equal to [2a.iv];
      (ii) for each year class i, determining number surviving into $i + 1$th class in following year (using equation 5, Table 24);
      (iii) adjusting next year's female age structure of unit by the probability that the unit underwent fission in current year (by equation 6, Table 24).
   (e) Repeat from [2a] for next cycle (year) until male's probability of retaining tenure to end of previous year [2b.iii] is zero.
3. Recommence at [2] for next age at start of tenure, with female composition of unit reset as specified in [1].

    The female age structures of the harems of different sizes used in the simulations are given in Table A.2. For reasons of space, a 2-year interval is used to classify the females in the table, although the simulations actually used 1-year intervals. The age structures were based on the typical compositions of units in the population.

C. Linear Model of the Gelada Socio-Ecological System

1. Specify required range in altitude and annual rainfall.
2. For each specified value of rainfall, calculate the annual birth rate (using equation 1, Table 34).

Table A.2
Age structure of harems used in the simulation of male reproductive strategies.

Harem Size	____				Female Age (years)					
	0–1	2–3	4–5	6–7	8–9	10–11	12–13	14–15	16–17	18–19
1	—	—	1	—	—	—	—	—	—	—
2	0.25	0.25	1	—	1	—	—	—	—	—
3	0.5	0.5	1	1	—	1	—	—	—	—
4	1	0.5	1	1	1	—	1	—	—	—
5	1	1	1.5	1	1	0.5	0.5	—	0.5	—
6	1.5	1	2	1.5	1	1	—	0.5	—	—
7	1.5	1.5	2	2	1	1	0.5	—	0.5	—
8	2	1.5	2	2	1.5	1	1	—	0.5	—
9	2.5	1.5	2.5	2	1.5	1.5	1	—	0.5	—
10	2.5	2	3	2	1.5	1.5	1.5	—	0.5	—

3. Adjust [2] for each specified value of altitude (using equation 2, Table 34, relative to a baseline at 3300 m asl; this baseline was defined as Sankaber since equation 1 was determined only from Sankaber data).
4. For each climatic regime specified in [3]:
   (a) calculate adult sex ratio (from [3] using equation 3, Table 34);
   (b) calculate percentage of harem-holding males (from [4a] using equation 4, Table 34);
   (c) calculate the percentage of multi-male units (from [4b] using equation 5, Table 34);
   (d) calculate mean harem size (from [4c] using equation 6, Table 34);
   (e) calculate:
      (i) probability of takeover (from [4d] using equation 7, Table 34);
      (ii) proportion of males attempting entry by takeover (from [4d] using equation 8, Table 34].

D. Dynamic Model of the Gelada Socio-Ecological System for Sankaber

I here give the outline for the complete model incorporating all the feedback loops and random bias effects listed in the text. In view of the feedback loops, the actual computer program is also given below.

1. Specify number of cycles (years) for which model will run.
2. Determine rainfall from *either*
   (a) a random number within specified limits,
      or
   (b) a defined pattern of specified cycle length.
3. Calculate gross annual birth rate (from [2] using equation 1, Table 34).
4. Adjust gross birth rate to take account of harem size effect (from [3] and [10], using equation 1, Table 39, relative to a base equivalent to the overall mean harem size observed in population—here 4.25 females).
5. Adjust net birth rate (from [4]) to take account of reproductive synchrony due to takeovers in current and previous years:
   (a) calculate proportion of females coming into oestrus prematurely in current year (half of each unit taken over: i.e. half of [11a]);
   (b) adjust net birth rate for current year (from [4] and [5a] using equation 2, Table 39).
6. Calculate small sample bias for neonatal sex ratio (from random number, using equation 3, Table 39).
7. Calculate adult sex ratio in the $(i + 6)$th year (from [6] using equation 3, Table 34, adjusted by equation 4, Table 39, relative to a base equivalent to a neonatal sex ratio of unity).
8. Calculate percentage of harem-holding males (from [7] using equation 4, Table 34).
9. Calculate percentage of multimale units (from [8] using equation 5, Table 34).
10. Calculate mean harem size (from [9] using equation 6, Table 34).
11. Calculate:
    (a) probability of takeover (from [10] using equation 7, Table 34);
    (b) proportion of males attempting entry by takeover (from [10] using equation 8, Table 34).
12. Repeat [2] for next cycle.

Computer program in BASIC

Variables: A - gross adult sex ratio
             B - gross annual birth rate
             C - net annual birth rate
             D - adjusted annual birth rate
             E - proportion of males attempting to enter units by
                 takeover
             H - percentage of harem-holding males
              I - cycle number
             K - Direction of neonatal sex ratio bias
             M - percentage of multimale units
             N - mean harem size
             P - proportion of females brought into oestrus prema-
                 turely due to harem takeover
             Q - the value of P in the previous cycle
             R - annual rainfall (mm)
             S - adjusted adult sex ratio
             T - probability of takeover per unit
             V - duration of simulation (number of cycles)
             Y - output matrix for birth rate over I cycles

```
 10   REM Dynamic Model of Gelada Socio-Ecological System
 20   REM (random rainfall model)
 30   DIN Y(V,1)
 40   FOR I = 1 TO V
 50   LET R = 1200 + 200*RND
 60   LET B = 1.3541 − 7.20000E−04*R
 70   IF I < 6 THEN D = B \ GO TO 120
 80   LET C = B*((.5855 − .0238*Y(I−6, 1))/.4903
 90   LET P =.5*(.1455*Y(I−6, 1) − .4371)
100   LET D = P + C*(1 − P − Q)
110   LET Q = P
120   IF RND <.5 THEN K =.1193 \ GO TO 140
130   LET K = −.1193
140   LET A =.5028 + K*N
150   LET S = (1.0279 + 3.3197*D)*((10.0163 − 13.4989*A)/3.2669)
160   LET H = 17.521 + 17.542*S
170   LET M = 105.864 − 1.133*H
180   IF M < 0 THEN M = 0
190   LET N = 3.6249 + .0157*M
200   LET T =.1945*Y(I−6, 1) − .3652
```

```
210   IF T < 0 THEN T = 0
220   LET E = .1455*Y(I − 6, 1) − .4371
230   IF E < 0 THEN E = 0
240   LET Y(I,0) = D
250   PRINT I,R,D,S,N,H,M,T,E
260   NEXT I
270   END
```

References

Altmann, J. (1974). Observational study of behaviour: sampling methods. *Behaviour 49:* 227–267.

Altmann, J. (1980). *Baboon Mothers and Infants*. Harvard University Press, Cambridge, Mass.

Altmann, S.A. (1974). Baboons, space, time, and energy. *Am. Zool. 14:* 221–240.

Altmann, S.A. and Altmann, J. (1979). Demographic constraints on behavior and social organization. In: *Primate Ecology and Human Evolution* (ed. I. Bernstein and E.O. Smith), pp. 47–64. Garland STMP, New York.

Alvarez, F. (1973). Periodic changes in the bare skin areas of *Theropithecus gelada. Primates 14:* 195–199.

Anderson, C.O. and Mason, W.A. (1978). Competitive social strategies in groups of deprived and experienced rhesus monkeys. *Dev. Psychobiol. 11:* 289–299.

Angst, W. and Thommen, D. (1977). New data and a discussion of infant killing in Old World monkeys and apes. *Folia primatol. 27:* 198–229.

Belonje, P.C. and van Niekerk, C.H. (1975). A review of the influence of nutrition upon the oestrous cycle and early pregnancy in the mare. *J. Reprod. Fertil., Suppl. 23:* 167–169.

Berryman, A.A. (1981). *Population Systems: A General Introduction.* Plenum Press, New York.

Bertram, B.C.R. (1975). Social factors influencing reproduction in wild lions. *J. Zool. 177:* 463–482.

Bertram, B.C.R. (1982). Problems with altruism. In: *Current Problems in Sociobiology* (ed. King's College Sociobiology Group), pp. 251–267. Cambridge University Press, Cambridge.

Bowman, L.A., Dilley, S.R., and Keverne, E.B. (1978). Suppression of oestrogen-induced LH surges by social subordination in talapoin monkeys. *Nature 275:* 56–58.

Bramblett, C. (1970). Coalitions among gelada baboons. *Primates 11:* 327–333.

Busse, C. and Hamilton, W.J. (1981). Infant carrying by male chacma baboons. *Science 212:* 1281–1283.

Buzas, M.A. and Gibson, T.G. (1969). Species diversity: benthonic Forminifera in the western North Atlantic. *Science 163:* 72–75.

Calow, P. and Townsend, C.R. (1981). Energetics, ecology and evolution. In: *Physiological Ecology* (ed. C.R. Townsend and P. Calow), pp. 3–19. Blackwell, Oxford.

Caughley, G. (1977). *Analysis of Vertebrate Populations*. Wiley, Chichester.

Chapman, M. and Hausfater, G. (1979). The reproductive consequences of infanticide in langurs: a mathematical model. *Behav. Ecol. Sociobiol. 5:* 227–240.

Charlesworth, B. (1980). *Evolution in Age-Structured Populations*. Cambridge University Press, Cambridge.

Clutton-Brock, T.H. (1972). "Feeding and Ranging Behaviour of the Red Colobus Monkey." Ph.D. thesis, University of Cambridge.

Clutton-Brock, T.H. (1982). Sons and daughters. *Nature 298:* 11–13.

Clutton-Brock, T.H. and Albon, S.D. (1982). Parental investment in male and female offspring in mammals. In: *Current Problems in Sociobiology* (ed. King's College Sociobiology Group), pp. 223–247. Cambridge University Press, Cambridge.

Crook, J.H. (1966). Gelada baboon herd structure and movement: a comparative report. *Symp. Zool. Soc. Lond. 18:* 237–258.

Crook, J.H. (1970). Socio-ecology of primates. In: *Social Behaviour in Birds andMammals* (ed. J.H. Crook), pp. 103–166. Academic Press, London.

Crook, J.H. (1980). *The Evolution of Human Consciousness*. Clarendon Press, Oxford.

Crook, J.H. (1983). On attributing consciousness to animals. *Nature 303:* 11–14.

Crook, J.H. and Aldrich-Blake, F.P.G. (1968). Ecological and behavioural contrasts between sympatric ground-dwelling primates in Ethiopia. *Folia primatol. 8:* 192–227.

Crook, J.H. and Garlan, J.S. (1966). Evolution of primate societies. *Nature 210:* 1200–1203.

Daly, M. and Wilson, M. (1983). *Sex, Evolution and Behavior*. 2nd edition. Willard Grant Press, Boston.

Datta, S.B. (1981). "Dynamics of Dominance among Free-Ranging Rhesus Females." Ph.D. Thesis, University of Cambridge.

Davies, N.B. (1982). Behaviour and competition for scarce resources. In: *Current Problems in Sociobiology* (ed. King's College Sociobiology Group), pp. 363–380. Cambridge University Press, Cambridge.

Dawkins, R. (1976). *The Selfish Gene*. Oxford University Press, Oxford.

Dawkins, R. (1980). Good strategy or evolutionarily stable strategy?

In: *Sociobiology: Beyond Nature/Nurture?* (ed. G.W. Barlow and J. Silverberg), pp. 331–367. Westview Press, Boulder.

Denham, W.W. (1971). Energy relations and some basic properties of primate social organization. *Amer. Anthrop. 73:* 77–95.

Dittus, W.P.J. (1975). Population dynamics of the toque monkey, *Macaca sinica.* In: *Socioecology and Psychology of Primates* (ed. R.H. Tuttle), pp. 125–151. Mouton, The Hague.

Dunbar, R.I.M. (1976). Australopithecine diet based on a baboon analogy. *J. Human Evol. 5:* 161–167.

Dunbar, R.I.M. (1977a). Feeding ecology of gelada baboons: a preliminary report. In: *Primate Ecology* (ed. T.H. Clutton-Brock), pp. 251–273. Academic Press, London.

Dunbar, R.I.M. (1977b). Age-dependent changes in sexual skin colour and associated phenomena of female gelada baboons. *J. Human Evol. 6:* 667–672.

Dunbar, R.I.M. (1977c). The gelada baboon: status and conservation. In: *Primate Conservation* (ed. Rainier of Monaco and G.H. Bourne), pp. 363–383. Academic Press, New York.

Dunbar, R.I.M. (1978a). Competition and niche separation in a high altitude herbivore community in Ethiopia. *E. Afr. Wildl. J. 16:* 183–199.

Dunbar, R.I.M. (1978b). Sexual behaviour and social relationships among gelada baboons. *Anim. Behav. 26:* 167–178.

Dunbar, R.I.M. (1979a). Population demography, social organization and mating strategies. In: *Primate Ecology and Human Evolution* (ed. I. Bernstein and E.O. Smith), pp. 65–88. Garland STMP, New York.

Dunbar, R.I.M. (1979b). Structure of gelada baboon reproductive units. I. Stability of social relationships. *Behaviour 69:* 72–87.

Dunbar, R.I.M. (1980a). Demographic and life history variables of a population of gelada baboons (*Theropithecus gelada*). *J. Anim. Ecol. 49:* 485–506.

Dunbar, R.I.M. (1980b). Determinants and evolutionary consequences of dominance among female gelada baboons. *Behav. Ecol. Sociobiol. 7:* 253–265.

Dunbar, R.I.M. (1982a). Intraspecific variations in mating strategy. In: *Perspectives in Ethology*, Vol. 5 (ed. P. Klopfer and P. Bateson), pp. 385–431. Plenum Press, New York.

Dunbar, R.I.M. (1982b). Adaptation, fitness and the evolutionary tautology. In: *Current Problems in Sociobiology* (ed. King's College Sociobiology Group), pp. 9–28. Cambridge University Press, Cambridge.

Dunbar, R.I.M. (1982c). Structure of social relationships in a captive group of gelada baboons: a test of some hypotheses derived from studies of a wild population. *Primates 23:* 89–94.

Dunbar, R.I.M. (1982d). Review of "Animal Behaviour: It's Development, Ecology and Evolution" (R.A. Wallace). *Behav. Processes 7:* 190–192.

Dunbar, R.I.M. (1982e). Sociobiology, biosophy and human behaviour. *Ethol. Sociobiol. 2:* 187–190.

Dunbar, R.I.M. (1983a). Life-history tactics and alternative strategies of reproduction. In: *Mate Choice* (ed. P.P.G. Bateson), pp. 423–433. Cambridge University Press, Cambridge.

Dunbar, R.I.M. (1983b). Structure of gelada baboon reproductive units. II. Social relationships between reproductive females. *Anim. Behav. 31:* 556–564.

Dunbar, R.I.M. (1983c). Structure of gelada baboon reproductive units. III. The male's relationship with his females. *Anim. Behav. 31:* 565–575.

Dunbar, R.I.M. (1983d). Structure of gelada baboon reproductive units. IV. Integration at group level. *Z. Tierpsychol. 63:* 265–282.

Dunbar, R.I.M. (1983e). Relationships and social structure in gelada and hamadryas baboons. In: *Primate Social Relationships* (ed. R. Hinde), pp. 229–307. Blackwell, Oxford.

Dunbar, R.I.M. (1984a). Theropithecines and hominids: contrasting solutions to the same ecological problem. *J. Human Evol. 12:* 647–658.

Dunbar, R.I.M. (1984b). Use of infants by male gelada in agonistic contexts: agonistic buffering, progeny protection or soliciting support? *Primates 25:* 28–35.

Dunbar, R.I.M. and Crook, J.H. (1975). Aggression and dominance in the weaver bird, *Quelea quelea. Anim Behav. 23:* 450–459.

Dunbar, R.I.M. and Dunbar, P. (1974a). Ecological relations and niche separation between sympatric terrestrial primates in Ethiopia. *Folia primatol. 21:* 36–60.

Dunbar, R.I.M. and Dunbar, P. (1974b). Ecology and population dynamics of *Colobus guereza* in Ethiopia. *Folia primatol. 21:* 188–208.

Dunbar, R.I.M. and Dunbar, P. (1974c). The reproductive cycle of the gelada baboon. *Anim. Behav. 22:* 203–210.

Dunbar, R.I.M. and Dunbar, P. (1975). *Social Dynamics of Gelada Baboons.* Karger, Basel.

Dunbar, R.I.M. and Dunbar, P. (1976). Contrasts in social structure among black-and-white colobus monkeys. *Anim. Behav. 24:* 84–92.

Dunbar, R.I.M. and Dunbar, P. (1977). Dominance and reproductive success among female gelada baboons. *Nature 266:* 351–352.

Dunbar, R.I.M. and Dunbar, P. (1981). The grouping behaviour of male walia ibex with special reference to the rut. *Afr. J. Ecol. 19:* 251–263.

Dunbar, R.I.M. and Sharman, M. (1983). Is social grooming altruistic? *Z. Tierpsychol. 64:* 163–173.

Eisenberg, J.F., Dittus, W.P.J., Fleming, T.H., Green, K., Struhsaker, T.T., and Thorington, R.W. (1981). *Techniques for the Study of Primate Population Ecology.* National Academy Press, Washington, D.C.

Fédération CECOS, Schwartz, D., and Mayeaux, M.J. (1982). Female fecundity as a function of age. *New England J. Med. 306:* 404–406.

Fedigan, L.M. (1972). Roles and activities of male geladas (*Theropithecus gelada*). *Behaviour 46:* 83–90.

Feyerabend, P.K. (1963). How to be a good empiricist—a plea for tolerance in matters epistemological. In: *Philosophy of Science, the Delaware Seminar*, Vol. 2 (ed. B. Baumrin), pp. 3–39. Interscience Publishers, New York.

Fisher, R.A. (1930). *The Genetical Theory of Natural Selection.* Clarendon Press, Oxford.

Frame, L.H. and Frame, G.W. (1977). Female African wild dogs emigrate. *Nature 263:* 227–229.

Fretwell, S.D. (1972). *Populations in a Seasonal Environment.* Princeton University Press, Princeton, N.J.

Fuchs, S. (1982). Optimality of parental investment: the influence of nursing on reproductive success of mother and female young house mice. *Behav. Ecol. Sociobiol. 10:* 39–51.

Gadgil, M. (1972). Male dimorphism as a consequence of sexual selection. *Amer. Natur. 106:* 574–580.

Gale, G. (1979). *Theory of Science.* McGraw-Hill, New York.

Gardiner, P. (1952). *The Nature of Historical Explanation.* Oxford University Press, Oxford.

Geach, P. (1975). Teleological explanation. In: *Explanation* (ed. S. Korner), pp. 76–95. Blackwell, Oxford.

Geist, V. (1971). *Mountain Sheep.* University of Chicago Press, Chicago.

Gibson, R.M. and Guinness, F.E. (1980). Differential reproduction among red deer (*Cervus elaphus*) stags on Rhum. *J. Anim. Ecol. 49:* 199–208.

Ginsburg, B. and Allee, W.C. (1942). Some effects of conditioning on social dominance and subordination in inbred strains of mice. *Physiol. Zool. 15:* 485–506.

Goodman, L.A. (1968). The analysis of cross-classified data: independence, quasi-independence and interactions in contingency tables with or without missing entries. *J. Amer. Stat. Assoc. 63:* 1091–1131.

Gosling, M. (1974). The social behaviour of Coke's hartebeest (*Alcelaphus buselaphus cokei)*. In: *The Behaviour of Ungulates and Its Relation to Management* (ed. V. Geist and F.R. Walther), pp. 488–511. IUCN, Morges.

Gosling, L.M. (1981). Climatic determinants of spring littering by feral coypus, *Myocastor coypus. J. Zool. 195:* 281–288.

Goss-Custard, J.D., Dunbar, R.I.M., and Aldrich-Blake, F.P.G. (1972). Survival, mating and rearing strategies in the evolution of primate social structure. *Folia primatol. 17:* 1–19.

Gouzoules, H., Gouzoules, S., and Fedigan, L. (1982). Behavioural dominance and reproductive success in female Japanese monkeys *(Macaca mulatta). Anim. Behav. 30:* 1138–1150.

Grafen, A. (1982). How not to measure inclusive fitness. *Nature 298:* 425–426.

Graham, C.E., Kling, O.R., Steiner, R.A. (1979). Reproductive senescence in female non-human primates. In: *Ageing in Non-human Primates* (ed. D.M. Bowden), pp. 183–202. Van Nostrand Reinhold, New York.

Griffin, D.R. (1981). *The Question of Animal Awareness*, 2nd ed. Rockefeller University Press, New York.

Guhl, A.M. (1968). Social inertia and social stability in chickens. *Anim. Behav. 16:* 219–232.

Hall, K.R.L. (1965). Behaviour and ecology of the wild patas monkey, *Erythrocebus patas,* in Uganda. *J. Zool. 14:* 15–87.

Hamilton, W.D. (1964). The genetical evolution of social behaviour. I, II. *J. Theor. Biol. 7:* 1–52.

Harcourt, A.H. (1978). Strategies of emigration and transfer by primates, with particular reference to gorillas. *Z. Tierpsychol. 48:* 401–420.

Harcourt, A.H. and Stewart, K.J. (1981). Gorilla male relationships: can differences during immaturity lead to contrasting reproductive tactics in adulthood? *Anim. Behav. 29:* 206–210.

Harcourt, A.H., Harvey, P.H., Larson, S.G., and Short, R.V. (1981).

Testis weight, body weight and breeding system in primates. *Nature 293:* 55–57.

Hausfater, G. (1975). *Dominance and Reproduction in Baboons: A Quantitative Study.* Karger, Basel.

Hendrickx, A.G. and Kraemer, D.G. (1969). Observations on the menstrual cycle, optimal mating time and pre-implantation embryos of the baboon, *Papio anubis* and *Papio cynocephalus. J. Reprod. Fertil., Suppl. 6:* 119–128.

Hinde, R.A. (1975). Interactions, relationships and social structure in non-human primates. In: *Proceedings of the Fifth Congress of the International Primatological Society* (ed. S. Kondo, M. Kawai, A. Ehara, and S. Kawamura), pp. 13–24. Japan Science Press, Tokyo.

Hinde, R.A., ed. (1983). *Primate Social Relationships: An Integrated Approach.* Blackwell, Oxford.

Hirshfield, M.F. and Tinkle, D.W. (1975). Natural selection and the evolution of reproductive effort. *Proc. Nat. Acad. Sci. USA 72:* 2227–2231.

Howard, R.D. (1979). Estimating reproductive success in natural populations. *Amer. Natur. 114:* 221–231.

Hrdy, S.B. (1977). *The Langurs of Abu.* Harvard University Press, Cambridge, Mass.

Hrdy, S.B. (1979). Infanticide among animals: a review, classification and examination of the implications for the reproductive strategies of females. *Ethol. Sociobiol. 1:* 13–40.

Humphrey, N.K. (1976). The social function of intellect. In: *Growing Points in Ethology* (ed. P.P.G. Bateson and R.A. Hinde), pp. 303–317. Cambridge University Press, Cambridge.

Iwamoto, T. (1979). Feeding ecology. In: M. Kawai (1979a), pp. 279–330.

Iwamoto, T. and Dunbar, R.I.M. (1983). Thermoregulation, habitat quality and the behavioural ecology of gelada baboons. *J. Anim. Ecol. 53:* 357–366.

Janis, C. (1976). The evolutionary strategy of the Equidae and the origins of rumen and cecal digestion. *Evolution 30:* 757–774.

Jarman, M.V. (1979). *Impala Social Behaviour: Territory, Hierarchy, Mating and Use of Space.* Paul Parey, Berlin.

Jarman, M.V. and Jarman, P.J. (1973). Daily activity of impala. *E. Afr. Wildl. J. 11:* 75–92.

Jeffers, J.N.R. (1978). *An Introduction to Systems Analysis: With Ecological Applications.* Arnold, London.

Jolly, C.J. (1970a). The seed-eaters: a new model of hominid differentiation based on a baboon analogy. *Man 5:* 5–26.

Jolly, C.J. (1970b). The large African monkeys as an adaptive array. In: *Old World Monkeys* (ed. J. Napier and P. Napier), pp. 139–174. Academic Press, London.

Jolly, C.J. (1972). The classification and natural history of *Theropithecus (Simopithecus)* (Andrews, 1916), baboons of the African Plio-Pleistocene. *Bull. Brit. Mus. Nat. Hist. (Geol.) 22:* 1–123.

Kawai, M., ed. (1979a). *Ecological and Sociological Studies of Gelada Baboons.* Karger, Basel, and Kodansha, Tokyo.

Kawai, M. (1979b). Auditory communication and social relations. In: M. Kawai (1979a), pp. 219–241.

Kawai, M., Dunbar, R.I.M., Ohsawa, H., and Mori, U. (1983). Social organisation of gelada baboons: social units and definitions. *Primates 24:* 1–13.

Kawai, M. and Iwamoto, T. (1979). Nomadism and activities. In: Kawai, M. (1979a), pp. 251–278.

Kawai, M. and Mori, U. (1979). Spacing within units and unit integrity. In: Kawai, M. (1979a), pp. 199–217.

Kummer, H. (1968). *Social Organisation of Hamadryas Baboons.* Karger, Basel.

Kummer, H. (1975). Rules of dyad and group formation among captive gelada baboons (*Theropithecus gelada*). In: *Proceedings of the Fifth Congress of the International Primatological Society* (ed. S. Kondo, M. Kawai, A. Ehara, and S. Kawamura), pp. 129–159. Japan Science Press, Tokyo.

Kummer, H. (1978). On the value of social relationships to non-human primates: a heuristic scheme. *Soc. Sci. Inform. 17:* 687–705.

Kummer, H. (1982). Social knowledge in free-ranging primates. In: *Animal Mind—Human Mind* (ed. D.R. Griffin), pp. 113–150. Springer-Verlag, Berlin.

Kummer, H., Abegglen, J.J., Bachmann, C., Falett, J., and Sigg, H. (1978). Grooming relationship and object competition among hamadryas baboons. In: *Recent Advances in Primatology*, Vol. 1. *Behaviour* (ed. D.J. Chivers and J. Herbert), pp. 31–38. Academic Press, London.

Kurland, J.A. (1977). *Kin Selection in the Japanese Monkey*, Karger, Basel.

Lack, D. (1966). *Population Studies of Birds.* Clarendon Press, Oxford.

Laing, J.A. and Leech, F.B. (1975). The frequency of infertility in thoroughbred mares. *J. Reprod. Fertil., Suppl. 23:* 307–310.

LeBoeuf, B.J. (1974). Male-male competition and reproductive success in elephant seals. *Am. Zool. 14:* 163–176.

Lewis, M. (1959). *Amharic-to-English Transliteration System for Geographic Names and Terms*. Mapping and Geography Institute, Addis Ababa.

Lindley, D.V. (1970). *Introduction to Probability and Statistics from a Bayesian Viewpoint*. 2 vols. Cambridge University Press, Cambridge.

Logan, W.P.D. (1959). Vital statistics of reproduction. In: *British Obstetric and Gynaecological Practice*, Vol. 1. *Obstetrics* (ed. E. Holland), pp. 1172–1198. Heinemann, London.

Masui, K., Sugiyama, Y, Nishimura, A., and Ohsawa, H. (1975). The lifetable of Japanese monkeys at Takasakiyama. In: *Contemporary Primatology* (ed. M. Kawai, S. Kondo, and A. Ehara), pp. 401–406. Karger, Basel.

Maxwell, A.E. (1968). *Analysing Qualitative Data*. Methuen, London.

May, R.M. (1976). Simple mathematical models with very complicated dynamics. *Nature 261:* 459–467.

May, R.M. and Oster G.F. (1976). Bifurcations and dynamic complexity in simple ecological models. *Amer. Natur. 110:* 573–599.

Maynard Smith, J. (1974). The theory of games and the evolution of animal conflict. *J. Theor. Biol. 47:* 209–221.

Maynard Smith, J. (1977). Parental investment: a prospective analysis. *Anim. Behav. 25:* 1–9.

Maynard Smith, J. (1978). Optimisation theory in evolution. *Ann. Rev. Ecol. Syst. 9:* 31–56.

Maynard Smith, J. and Price G.R. (1973). The logic of animal conflict. *Nature 246:* 15–18.

McFarland, D. and Houston, A. (1981). *Quantitative Ethology: A State Space Approach*. Pitman, London.

Morgan, B.J.T., Simpson, M.J.A., Hanby, J.P., and Hall-Craggs, J. (1976). Visualising interaction and sequential data in animal behaviour: theory and application of cluster-analysis methods. *Behaviour 56:* 1–43.

Mori, A. (1977). The social organisation of the provisioned Japanese monkey troops which have extraordinarily large population sizes. *J. Anthrop. Soc. Nippon 85:* 325–345.

Mori, U. (1979a). Reproductive behaviour. In: M. Kawai (1979a), pp. 183–197.

Mori, U. (1979b). Individual relationships within a unit. In: M. Kawai (1979a), pp. 94–124.

Mori, U. (1979c). Unit formation and the emergence of a new leader. In: M. Kawai (1979a), pp. 156–181.

Mori, U. (1979d). The development of sociability and social status. In: M. Kawai (1979a), pp. 125–154.

Mori, U. and Dunbar, R.I.M. (1984). Changes in the reproductive condition of female gelada baboons following the takeover of one-male units. *Z. Tierpsychol.* (in press).

Nagel, U. (1973). A comparison of anubis baboons, hamadryas baboons and their hybrids at a species border in Ethiopia. *Folia primatol. 19:* 104–165.

Nagel, U. (1979). On describing primate groups as systems: the concept of ecosocial behaviour. In: *Primate Ecology and Human Evolution* (ed. I.R. Bernstein and E.O. Smith), pp. 313–339. Garland STMP, New York.

Napier, J.R. and Napier, P. (1967). *A Handbook of Living Primates*. Academic Press, London.

Ohsawa, H. (1979). The local population and environment of the Gich area. In: M. Kawai, (1979a), pp. 3–45.

Ohsawa, H. and Dunbar, R.I.M. (1984). Variations in the demographic structure and dynamics of gelada baboon populations. *Behav. Ecol. Sociobiol.* (in press).

Packer, C. (1979a). Male dominance and reproductive activity in *Papio anubis. Anim. Behav. 17:* 37–45.

Packer, C. (1979b). Inter-troop transfer and inbreeding avoidance in *Papio anubis. Anim. Behav. 27:* 1–36.

Parker, G.A. (1974). Assessment strategy and the evolution of fighting behaviour. *J. Theor. Biol. 47:* 157–184.

Parker, G.A. (1978). Selfish genes, evolutionary games and the adaptiveness of behaviour. *Nature 274:* 849–855.

Parker, G.A. and Rubenstein, D.I. (1981). Role assessment, reserve strategy and acquisition of information in asymmetric animal conflicts. *Anim. Behav. 29:* 221–240.

Pearson, R. (1978). *Climate and Evolution*. Academic Press, London.

Peet, R.K. (1974). The measurement of species diversity. *Ann. Rev. Ecol. Syst. 5:* 285–307.

Platt, J.R. (1964). Strong inference. *Science 146:* 347–353.

Platt, T. and Denman, K.L. (1975). Spectral analysis in ecology. *Ann. Rev. Ecol. Syst. 6:* 189–210.

Pollard, J.H. (1977). *A Handbook of Numerical and Statistical Techniques*. Cambridge University Press, Cambridge.

Popp, J.L. and DeVore, I. (1980). Aggressive competition and social dominance theory: synopsis. In: *The Great Apes* (ed. D.A. Hamburg and E.R. McCown), pp. 317–338. Benjamin / Cummings, Menlo Park, Calif.

Portman, O.W. (1970). Nutritional requirements (NRC) of nonhuman primates. In: *Feeding and Nutrition of Nonhuman Primates* (ed. R.S. Harris), pp. 87–115. Academic Press, New York.

Raemakers, J. (1979). Ecology of sympatric gibbons. *Folia primatol. 31:* 227–245.

Rapoport, A. and Chammar, A. (1965). *Prisoner's Dilemma.* University of Michigan Press, Ann Arbor, Mich.

Richard, A. (1970). A comparative study of the activity patterns and behaviour of *Alouatta villosa* and *Ateles geoffroyi. Folia primatol. 12:* 241–263.

Richman, B. (1976). Some vocal distinctive features used by gelada monkeys. *J. Acoust. Soc. Am. 60:* 718–724.

Richman, B. (1978). The synchronisation of voices by gelada monkeys. *Primates 19:* 569–581.

Riechert, S.E. (1978). Games spiders play: behavioural variability in territorial disputes. *Behav. Ecol. Sociobiol. 3:* 135–162.

Rowell, T.E. (1966). Hierarchy in the organisation of a captive baboon group. *Anim. Behav. 14:* 430–443.

Rowell, T.E. (1972). Female reproduction cycles and social behaviour in primates. *Adv. Study Behav. 4:* 69–105.

Rowell, T.E. (1978). How female reproduction cycles affect interaction patterns in groups of patas monkeys. In: *Recent Advances in Primatology,* Vol. 1. *Behaviour* (ed. D.J. Chivers and J. Herbert), pp. 489–490. Academic Press, London.

Rubenstein, D.I. (1981). Population density, resource patterning and territoriality in the Everglades pygmy sunfish. *Anim. Behav. 29:* 155–172.

Rubenstein, D.I. (1982). Risk, uncertainty and evolutionary strategies. In: *Current Problems in Sociobiology* (ed. King's College Sociobiology Group), pp. 91–111. Cambridge University Press, Cambridge.

Sade, D.S. (1972). Sociometrics of *Macaca mulatta.* I. Linkages and cliques in grooming matrices. *Folia primatol. 18:* 196–223.

Sade, D.S., Cushing, K., Cushing, P., Dunaid, J., Figueroa, A., Kaplan, J.R., Lauer, C., Rhodes, D., and Schneider, J. (1978). Population dynamics in relation to social structure on Cayo Santiago. *Yrbk. Phys. Anthrop. 20:* 253–262.

Sadleir, R.M.F.S. (1969). *The Ecology of Reproduction in Wild and Domestic Mammals.* Methuen, London.

Salmon, W.C. (1966). *The Foundations of Scientific Inference.* Pittsburgh University Press, Pittsburgh.

Schaffer, W.M. (1974). Optimal reproductive effort in fluctuating environments. *Amer. Natur. 108:* 783–790.

Schaller, G. (1977). *Mountain Monarchs.* University of Chicago Press, Chicago.

Schulman, S.R. and Chapais, B. (1980). Reproductive value and rank relations among macaque sisters. *Amer. Natur. 115:* 580–593.

Seyfarth, R.M. (1977). A model of social grooming among adult female monkeys. *J. Theor. Biol. 65:* 671–698.

Seyfarth, R.M. (1980). The distribution of grooming and related behaviours among adult female vervet monkeys. *Anim. Behav. 28:* 798–813.

Sharman, M. (1981). "Feeding, Ranging and Social Organisation of the Guinea baboon, *Papio papio."* Ph.D. thesis, University of St. Andrews.

Sharman, M. and Dunbar, R.I.M. (1982). Observer bias in selection of study group in baboon field studies. *Primates 23:* 567–573.

Shotake, T. (1980). Genetic variability within and between herds of gelada baboons in central Ethiopian highland. *Anthrop. Contemporanea 3:* 270.

Sibley, R.M. (1981). Strategies of digestion and defecation. In: *Physiological Ecology* (ed. C.R. Townsend and P. Calow), pp. 109–142. Blackwell, Oxford.

Siegel, S. (1956). *Nonparametric Statistics for the Behavioural Sciences.* McGraw-Hill, Toronto.

Sigg, H., Stolba, A., Abegglen, J–J., and Dasser, V. (1982). Life history of hamadryas baboons: physical development, infant mortality, reproductive parameters and family relationships. *Primates 23:* 473–487.

Silk, J.B. (1982). Altruism among female *Macaca radiata:* explanations and analysis of patterns of grooming and coalition formation. *Behaviour 79:* 162–188.

Simpson, M.J.A., Simpson, A.E., Hooley, J., and Zunz, M. (1981). Infant-related influences on inter-birth intervals in rhesus monkeys. *Nature 290:* 49–51.

Snowdon, C.T., Brown, C.H., and Petersen, M.R., eds. (1982). *Primate Communication.* Cambridge University Press, Cambridge.

Sokal, R.R. and Rolf, F.J. (1969). *Biometry: The Principles and Practice of Statistics in Biological Research.* Freeman, San Francisco.

Spivak, H. (1971). Ausdrucksformen und soziale Beziehungen in einer Dschelada-Gruppe (*Theropithecus gelada*) im Zoo. *Z. Tierpsychol. 28:* 279–296.

Stahli, P. and Zurbuchen, M. (1978). Two topographic maps 1:25000 of Simen, Ethiopia. In: *Simen Mountains—Ethiopia*, Vol. I. *Cartography and Its Application for Geographical and Ecological Problems* (ed. M. Messerli and K. Aerni), pp. 11–31. Geographisches Institut, University of Bern, Switzerland.

Stearns, S.C. (1976). Life-history tactics: a review of ideas. *Q. Rev. Biol. 51:* 3–47.

Stolba, A. (1979). "Entscheidungsfindung in Verbanden von *Papio hamadryas.*" Ph.D. thesis, University of Zurich.

Struhsaker, T.T. (1977). Infanticide and social organisation in the redtail monkey (*Cercopithecus ascanius schmidti*) in the Kibale Forest, Uganda. *Z. Tierpsychol. 45:* 75–84.

Strum, S.C. (1982). Agonistic dominance in male baboons: an alternate view. *Int. J. Primat. 3:* 175–202.

Sugiyama, Y. (1965). On the social change of hanuman langurs (*Presbytis entellus*) in their natural condition. *Primates 6:* 381–418.

Szalay, F.S. and Delson, E. (1979). *Evolutionary History of the Primates.* Academic Press, New York.

Teleki, G., Hunt, E.E., Pfifferling, J.H. (1976). Demographic observations (1963–1973) on the chimpanzees of Gombe National Park, Tanzania. *J. Human Evol. 5:* 559–598.

Thierauf, R.J. and Klemp, R.C. (1975). *Decision Making Through Operations Research.* 2nd edition. Wiley, New York.

Townsend, C.R. and Hughes, R.N. (1981). Maximising net energy returns from foraging. In: *Physiological Ecology* (ed. C.R. Townsend and P. Calow), pp. 86–108. Blackwell, Oxford.

Trivers, R.L. and Willard, D.E. (1973). Natural selection of parental ability to vary the sex ratio. *Science 179:* 90–92.

Walker Leonard, J. (1979). A strategy approach to the study of primate dominance behaviour. *Behav. Processes 4:* 155–172.

Wallace, R.A. (1980). *Animal Behaviour: Its Development, Ecology and Evolution.* Goodyear, Santa Monica, Calif.

Walters, J. (1981). Inferring kinship from behaviour: maternity determinations in yellow baboons. *Anim. Behav. 29:* 126–136.

Waser, P.M. (1977). Feeding, ranging and group size in the mangabey, *Cercocebus albigena.* In: *Primate Ecology* (ed. T.H. Clutton-Brock), pp. 183–222. Academic Press, London.

Wilson, E.O. (1975). *Sociobiology: The New Synthesis.* Harvard University Press, Cambridge, Mass.

Winkless, N. and Browning I. (1975). *Climate and the Affairs of Men.* Peter Davies, London.

Wood, C.A. and Lovett, R.R. (1974). Rainfall, drought and the solar cycle. *Nature 251:* 594–596.

Wrangham, R.W. (1980). Bipedal locomotion as a feeding adaptation in gelada baboons, and its implications for hominid evolution. *J. Human Evol. 9:* 329–331.

Wrangham, R.W. (1981). An ecological model of female-bonded primate groups. *Behaviour 76:* 262–300.

Author Index

Subject Index

Library of Congress Cataloging in Publication Data

Dunbar, R. I. M. (Robin Ian MacDonald), 1947–
  Reproductive decisions.

  (Monographs in behavior and ecology)
  Bibliography: p.
  Includes index.
  1. Gelada baboon—Reproduction. 2. Gelada
baboon—Behavior. 3. Social behavior in animals.
4. Mammals—Reproduction. 5. Mammals—Behavior.
I. Title. II. Series.
QL737.P93D76 1984   599.8'2   84-42584
ISBN 0-691-08360-6 (alk. paper)
ISBN 0-691-08361-4 (pbk.)